FREE FOR ALL

All around us – in fields, woods, hedgerows, along roadsides and on waste ground – there grow wild plants free for the picking; many of them highly nutritious and some with specific medicinal uses. In this book, the author – lifelong naturalist, broadcaster and contributor to *The Times* – describes about 150 so-called weeds; telling how to recognize them, where and when to find them, and how to prepare, cook and store them.

FREE FOR ALL

Weeds and Wild Plants as a Source of Food

by

CERES

Drawings by Alison Ross

THORSONS PUBLISHERS LIMITED
Wellingborough, Northamptonshire

First published 1977

ISBN 0 7225 0366 0 (hardback)

ISBN 0 7225 0349 0 (paperback)

Typeset by Specialised Offset Services Ltd., Liverpool
Printed and bound in Great Britain

CONTENTS

For E., B.C., D., J., L. and P.

INTRODUCTION

As the earth's natural covering, especially in such a temperate country as Britain and some northern parts of Europe, consists of a succession of wild plants, or 'weeds' as we frequently call those that are uncultivated, there is a rich source of free-for-all potential food plants available to us, if we know what to look for.

This book describes about 150 wild plants, with advice on culinary and medicinal uses, and the lesser known plants are illustrated to aid recognition. The availability of free food plants naturally varies according to where they can grow. In towns and cities where uncultivated ground is very scarce, plants are scarce too. However, even most urban areas have a few waste corners and council rubbish-tips, and if permission can be obtained to visit these, a great number of useful 'food' plants can be found.

Indeed some rubbish-tips nurture fine crops of potatoes, tomatoes, marrows, gourds, and so on, from the seed which has been chucked out of household dustbins. Botanists have found exotic weeds which they believe may have grown from the seed cleared out of the bottom of bird cages.

But apart from these casuals, most rubbish-tips also have good growths of Nettles, Sorrels, Dungweed, Chickweed and even Landcress. It is a pity that Bracken, that ubiquitous coverer of so much uncultivated ground, is not recommendably edible. Until recently its young shoots were thought to make a tasty free vegetable, but the fern is not included here as a fact has emerged from modern pharmacological research which suggests that Bracken contains a cancer-inducing element, or carcinogen.

In the old days, when weeds were used for nutrition and for medical and other purposes and people were far more knowledgeable about the countryside, most villages had their own 'herb-women' and there was seldom any worry about the identification of harmful or harmless plants.

Now we have travelled far from that stage and there is difficulty in being sure about what you are picking. There are plenty of useful wild flower books to consult if you cannot find an expert to help you. Collins' *Wild Flowers in Britain and Northern Europe* by R. and A. Fitter, with delightful pictures by Marjorie Blamey, is low-priced and a helpful guide to identification.

I have included a section about Poisonous Plants in this book, all of which must obviously be avoided.

In many ways toadstools can be more dangerous than flowering plants, for some of the most lethal fungi are not superficially unlike others that are harmless, and indeed the latter are sometimes sought after by gastronomes. There are identifying experts in a lot of Continental markets, but unfortunately not in Britain, although local naturalists or museums may be helpful. A useful book is Collins' *Guide to Mushrooms and Toadstools* by Hora and Lange, which has excellent pictures, but the characteristics of the most dangerous species should also be learnt carefully.

I do not know that anyone has yet written a book about the edible Seaweeds that can be found round our coasts or whether those of us who enjoy their benefits, which include their unusual flavours and excellent nutritious value, have been lucky in our identification! However, all the Seaweeds recommended in this book have been eaten through the ages by people in normal health, living near the sea so that they could use them while they were fresh. DO NOT BE TEMPTED to experiment with any others or with any plants unless you are certain of their identity.

Now that our population is so high and our food needs so great there are fewer and fewer places left for wild plants to grow in or for wild creatures to survive. A few of the plants which our ancestors used to pick for eating have now become rarities: Cowslips (which were once used for wine-making); Bath Asparagus (*Ornithogalum pyrenaicum*); Sea Kale (*Crambe maritima*); or the Oyster Plant (*Mertensia maritima*) give an example or two. Now, however, in the interest of the conservation of our heritage, these are among the rare wild flowers that should no longer be picked.

Others, like the familiar Blackberries and wild Raspberries, Hips, Haws and edible nuts, should be picked with care for they are the wild birds' and animals' food as well as being increasingly available to us now that transport into the country is so much easier. *Don't be greedy* when you are taking free food: leave some for the birds and animals and remember that man is not the only creature on earth.

The Secret Life of Plants by Tompkins and Bird (published by Allen Lane in 1973) gave proof that plants can feel. Think about this too as you pick your free vegetables and salad plants – and your garden plants too, for that matter.

And, finally, as I have been writing this book during a very hot, unusually dry summer when everything is covered with dust which has made careful washing most essential, let me remind you that thorough washing of all free plant food is important. *It is also important to avoid any polluted areas* if you are picking wild plants for food.

Good luck with your free food, and remember always to try a little of anything that is new to you (and only pick a little, as well) before embarking on a big meal of it!

Wild Arum
(poisonous)

Deadly Nightshade
(Poisonous)

Laburnum
(Poisonous)

Fly Agaric
(Poisonous)

POISON

POISONOUS – BEWARE!

There are plenty of poisonous plants that are confusable with those that are edible, especially amongst the big botanical family of the Umbellifers.

Avoid ALL the **Water Dropworts** (*Oenanthe* spp.), (see page 16), as well as the rare **Cowbane** (*Cicuta virosa*), which invariably grows near or beside water.

Hemlock (*Conium maculatum*) is another very poisonous Umbellifer; it is tall, fine-leaved and has purple blotches, spots or freckles on its smooth stems. The whole plant smells strongly of mice.

BEWARE too, of **Fool's Parsley** (*Aethusa cynapium*), a common rather frail garden weed, usually under eighteen inches tall, which has obvious down-hanging green bracts under its lacy head of small, white flowers.

Never be tempted to try any of the **Buttercups** – even picking them gives some people blisters, – and the green **Hellebores** are still more toxic. Both the attractive **Columbine** (*Aquilegia* spp.) and the **Monkshood** are very poisonous and the latter, as shown by its Latin name, *Aconitum,* has been known and unfortunately used as a competent poisoner throughout the ages. A Northern plant, **Baneberry** (*Actaea spicata*), is poisonous enough to cause acute gastro-enteritis if accidentally eaten.

BEWARE of both **Buckthorn** trees, the **Common Buckthorn**, (*Rhamnus cathartica*), which was once used as a cruel purgative, and **Alder Buckthorn** (*Rhamnus frangula*). It is dangerous to eat **Mistletoe** especially the berries, and also the berries of **Daphne Mezereon,** which have been known to cause blindness in children who have merely handled and squashed them and then rubbed their eyes. **Spurge Laurel** (*Daphne laureola*) is another plant to avoid, and so are the **White Bryony** (*Bryonia alba*), with its colourful, matt-surfaced berries, and the **Black Bryony** (*Tamus communis*), with late summer garlands of shiny yellow, red and orange fruits. ALL CHILDREN SHOULD BE TAUGHT WHILE VERY YOUNG not to touch any berries, however tempting they may look. The spikes of **Wild Arum,** which look harmless enough when they show up in the hedgerows in tight clusters of bright berries, can be lethal.

Deadly Nightshade (*Atropa belladonna*), with its big, evil-looking, dingy purple flowers that are succeeded by large and glossy black berries, is far less common than the equally poisonous garden-weed **Black Nightshade** (*Solanum nigrum*), or the related **Woody Nightshade** or **Bittersweet** (*Solanum dulcamara*). **Green Nightshade** (*S. sarrachoides*), which is just as dangerous, is becoming increasingly common on rubbish-tips and should be avoided.

The black berries of the **Privets** – the **Garden Privet** (*Ligustrum ovalifolium*) and the **Wild Privet** (*L. vulgare*) – are both poisonous, as are the delightful-looking fruits of the **Spindle-Tree** (*Euonymus europaeus*). **Yews** are known by most people to be acutely dangerous. All parts of this tree are poisonous to humans and grazing horses and cattle, except, curiously enough, the luscious-looking, bright red, fleshy cases round the actual little dark green naked seed itself. **Butcher's Broom** (see page 34) berries can cause nasty symptoms, and so can all parts of the **Dog's Mercuries** (*Mercuralis perennis* and *M. annua*).

The most dangerous poisonous plant now found growing in woods, on rubbish-tips and waste places, as well as being cultivated in gardens, is the **Laburnum** (*Laburnum anagyroides*). All parts of this tree are very poisonous indeed, including the seed pods with which children sometimes play at 'shops', using them as peas.

All the **Spurges** (*Euphorbia* spp.) are dangerous. If their milky juices ever get into the eyes, wash out frequently with a mild tepid solution of bicarbonate of soda, and also go to the doctor.

The now-rare **Henbane** (*Hyoscyamus niger*), which also reeks of mice, must definitely be avoided as an unsafe plant, as must the increasingly common, handsome **Thorn-apple** (*Datura* spp.)

Foxgloves, (*Digitalis purpurea*), **Lilies-of-the-Valley** (*Convallaria majalis*), **Fritillaries** (*Fritillaria meleagris*),

Herb Paris (*Paris quadrifolia*), and **Meadow Saffron** (*Colchicum autumnale*) are all dangerous plants. Even **Bluebells** have been known to cause nasty symptoms when children have nibbled flowers, leaves or bulbs.

Several weeds can cause allergic reactions of alarming kinds in different people. Anti-histamine treatment usually relieves these quickly.

It is foolish to try out *any* unrecognized **Fungi**. The **Fly Agaric** is easily recognized by its brilliant scarlet and white-warted cap, but is probably not as lethal as its related very pale green **Death Cap**, or the near-white **Panther Cap**, both of which are known to be killers.

More than twenty other different kinds of toadstools, which may look harmless enough, can at least cause frightening and painful, if not fatal, symptoms and unless an expert is available to identify every specimen, as is sometimes the case in European markets, anything about which you are at all uncertain should not be touched.

In any cases of suspected plant food poisoning, keep the victim warm and quiet, try to find out what they have been eating, or even handling, and seek medical advice quickly.

EDIBLE AND USEFUL PLANTS

ACORNS   *Quercus robur*

The fruits of the Oak tree have often been used as human food and the browsing rights of oak woods, where animals picked up whatever they could find for food, have been zealously apportioned throughout our history.

A diet of nothing but acorns, or a surfeit of them, can cause poisoning in cattle, sheep and even pigs. After good years when the trees were heavy with acorns, the nuts were collected and kept to be used in the winter, mixed in with other fodder.

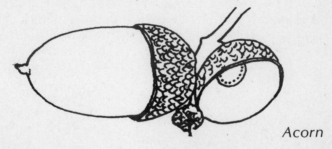

Acorn

Roasted acorns provide a coffee substitute for us which is said to be soothing 'to those with taut nerves'. When boiled and ground into flour, they were frequently, especially at famine times in this country, used to make bread, but they had to be soaked in a lye of hardwood ashes or charcoal first, to remove the tannic acid from them. Acorn and corn meal mixed together was said to make a bread that was more palatable than bread from acorn flour by itself.

A nineteenth-century American geologist, travelling through the Sacramento Valley, said that the Indians there lived primarily on a cake or bread made from acorn flour.

Chopped roasted acorns can be used instead of almonds.

ALEXANDERS   *Smyrnium olusatrum*

Country names: ASHINDER; SKIT; SKEET; HELLROOT.

This tall, early-flowering Umbelliferous plant which was originally brought to this country as a flavoursome vegetable or pot-herb, can now be found growing apparently wild on roadsides or path edges near the coast.

The leaves are a bright shiny green and their bases are broad, sometimes striped longitudinally with purple, and the plant is hairless. It smells of Angelica. The flower-heads are umbrella-shaped and dense with small yellow flowers.

BEWARE of confusing Alexanders with the equally vivid green leaves of **Hemlock Water Dropwort** which appears at the same time and grows in similar places but which is *poisonous*. Hemlock Water Dropwort smells much more rank, with a strong parsley-like undertone. It has less densely-packed umbels of white flowers. There is a little difference in the appearance of the leaves (see pictures below).

Alexanders   *Hemlock Water Dropwort*

Botanical experts should be sought by free-food seekers to identify these two plants before Alexanders is gathered for eating by the unknowledgeable. Once you can recognize it with certainty, it can either be cooked as a spring-green vegetable or used raw as a salad plant. The very young leaves are delicious for sandwich-fillers and the buds of the big flower-heads can also be chopped or 'frittered' like those of the Elder. The young juicy stems can be tried out cut into four – or five-inch lengths and cooked like Asparagus.

ALFALFA   *Medicago sativa*

Other names: PURPLE FIELD CLOVER; PURPLE MEDIC; CHILEAN CLOVER; LUCERNE.

Though not strictly a wild plant in Britain and Northern Europe, Alfalfa has been grown as a green forage crop for so long here that in places it now grows in a naturalized state. Look for it along roadside verges or on farmland outside the confines of its crop fields.

Alfalfa

Alfalfa shoots chopped small make good extra bulk in salads. They can be grown from seed all through the winter – indoors in jars with muslin 'lids' that can be rinsed out twice a day with fresh water – to be eaten whole as sprouters three or four days after they start germinating. They taste of young green peas and it is worth buying a small quantity of seed, if there is no free supply in your neighbourhood, both for sprouters and to start a patch of Alfalfa or Lucerne in the garden.

This plant is rich in vitamins A, C, D, E, and P, and was the first plant in which pharmacologists discovered the blood-clotting vitamin K.

Herbalists use Alfalfa as a tonic and mild stimulant. It makes a refreshing tea which has the old reputation of being helpful to lactating mothers as well as for those suffering from hyper-acidity and even stomach ulcers.

ANGELICA *Angelica sylvestris*

Country names: WILD ANGELICA; GHOST-KEX; KEWSIES.

It is the cultivated Angelica which is so often used for candying and which produces the vivid green strips of sugared stems for cake and sweet decorations. (For instructions on candying, see page 117). Wild Angelica can be used as well, but being slightly bitter in flavour it needs more care, and dye may have to be added to produce the striking green of the commercial product.

Wild Angelica is a magnificent plant which produces its wide umbrella-shaped flower-heads in late summer. The stems of these and even the flowers themselves may be tinged with pink, or with a dull purple.

It grows in marshy meadows and has hollow and sometimes sparsely downy stems, and may reach five or six feet in height.

Angelica

Angelica has long been used as a medicinal herb and a tea made from its leaves or young stems has a wind-dispelling or carminative use which is also said to destroy the desire for alcohol. But BEWARE for it must not be drunk by diabetics as it may cause an excess or increase of sugar in the urine. Its seeds, however, can be chewed (see Fennel, page 56) safely by anyone else.

Cultivated Angelica (*Angelica archangelica*) grows into an even stouter plant and usually has no purple staining on its stems. It sets viable seed and is self-perpetuating in some gardens; it can also be found as an escape on rubbish-tips, which are such good places for incipient free-fooders to hunt about on.

Cultivated Angelica is naturalized in some places in Britain, particularly beside rivers, and in Northern Europe it seems to prefer growing near the sea.

ASH   *Fraxinus excelsior*

The fruits of this tree are usually known as Ash 'keys' because of the way they hang in bunches which are green at first and turn brown after ripening.

These 'keys' are pleasant to eat when peeled, if picked as soon as they ripen. They can be pickled green, according to the diarist Evelyn, but have to be boiled first to get rid of their bitterness. In fact, after throwing out their first water, they should be boiled again and drained and then put into jars and covered with spice-flavoured vinegar.

Timber from Ash trees is excellent for burning; but never shelter under an Ash during a storm, because of the old and possibly true tradition:

Avoid an Ash,
It courts the flash.

BALM   *Melissa officinalis*

Country names: BEE BALM; LEMON BALM.

Balm can be classed as a weed here in Britain because it spreads so fast in some gardens that the owners are forever chucking roots of it out. These seem to flourish 'exceeding well' on lane-side verges, waste places and rubbish-tips, especially if they are damp and shady.

The heart-shaped leaves of this plant were taken in the Middle Ages to indicate that it was meant to be used as 'an hearbe of the heart', and modern herbalists, strangely enough, find that it can still be used as a heart-tonic or cordial.

In the days when plants were thought to show outward signs of their properties, as Balm does, all kinds of indications were taken as clues as to their proper use. Indeed, the Doctrine of Signatures, as this theorizing was called, has often been ridiculed by later botanists who could not believe that the presence of say, a yellow sap, showed that the plant could act as a cure for biliousness; or that one with twin root-tubers resembling testicles would have aphrodisiac powers.

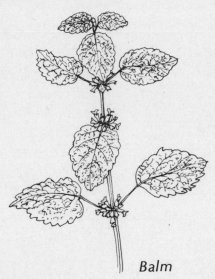

Balm

However, as is the case with Balm, the herbalists of the Middle Ages, whose knowledge was probably of ancient origin, are still being proved to be wiser than was hitherto suspected.

Balm can be used for flavouring sauces, stuffings and summer drinks, particularly claret-cup. The leaves can be chopped raw in salads and cooked with many different savoury dishes.

The fragrance is not released from the leaves until they are bruised, but honey-bees love the insignificant white flowers and seek them out. Beekeepers usually have this plant in their gardens, perhaps because of the old tradition which said that the bees regarded it as a homing plant which ensured that they would always come back to it.

BASIL   *Clinopodium vulgare*

Country names: WILD BASIL; COW BASIL; SWEET-LEAF.

This bright pink-flowered wild plant is related remotely to
the stronger-flavoured Mediterranean Sweet Basil
(*Ocimum basilicum*) and is commonly found in dry
places, particularly on chalk downs.

Wild Basil belongs to the same botanical family as the
Mints and its leaves can be used in all kinds of savoury
dishes as a pleasing flavourer.

I use it (or Marjoram leaves) in macaroni cheese and
enjoy the somewhat Clove-like taste.

In the old days before soap was a relatively cheap and
easily obtainable commodity, Wild Basil was distilled 'to
make sweet washing-waters'.

BEECH NUTS   *Fagus sylvatica*

Country name: BEECHMAST

The perfectly packed triangular fruits of the Beech tree
are only ripe when the green prickly outer fruit case
opens. They are very small once they have been peeled
but they have been appreciated as a wild food
throughout the ages, and 'Mankind', according to
Thomas Green, 'lived on beech nuts before the use of
corn, because it was the food of the common people'.

Both man and animals, including deer, eat these
triangular nuts and search avidly for them. Squirrels, field
mice and voles hide collections of them as stores for the
winter, sometimes biting out their embryonic roots and
shoots to prevent them from germinating.

Beech trees do not produce heavy crops of Beechmast
every year but only every two, three or four years. Their
fruits were deemed so important in Germany during
World War II that school children were given holidays in
order to collect them when they were plentiful.

In the early eighteenth century in Britain it was
suggested that our national debt should be paid off by
exporting 'beech nut oil' but this plan came to nothing.
As Beech nuts contain 42 per cent oil, and 22 per cent
protein and the valuable trace elements of calcium,

magnesium, phosphorus, potassium and silicon, however, they are always well worth looking for. They can be ground to make a coarse meal, or roasted and ground, like acorns, as a substitute for coffee.

BEEFSTEAK FUNGUS   *Fistulina hepatica*

Common names: BLEEDING-BRACKET-TOADSTOOL; *Langue de boeuf* (France).

When young, these fungi appear as fleshy outgrowths on the bark of some forest trees, particularly those of Oaks. They look very like curled-up ox-tongues at first and when cut their flesh resembles red, bleeding meat, but the taste, even after boiling or frying, is not comparable and *gourmets* find this a disappointing, rather acid, edible fungus, in spite of some seekers' recommendations.

BEET   *Beta vulgaris ssp. maritima*

Common names: SEA BEET; WILD SPINACH; SEA SPINACH.

The leaves of wild beet are one of the best of our free vegetables. The plant grows abundantly round the coast and produces crops of big, shiny, thick and sometimes red-veined, Spinach-like leaves from January until it comes into flower in about May or June.

I have been gathering them for years and have found that they are best cooked conservatively in very little water – just enough to cover the bottom of the pan – for only two or three minutes.

Sea Beet leaves are rich in vitamins A and C so the less cooking they have the better, but it is important that they are cooked long enough to be tender. I never bother to sieve them before serving but strain them off and dab a pat of butter on top of the steaming dishful. They are delicious with a poached egg on top and covered with a sprinkling of cheese, or chopped (cooked) sparsely into scrambled eggs. I also use them chopped with any remainders of ham as a filling for omelettes – a flavouring of cinnamon also makes a pleasing change.

During World War II the beaches were barricaded off and the supply of Sea Spinach was cut down badly because most of it was inaccessible. However, the plant was found to grow in the estuaries and a little way up the banks of some tidal rivers, and the leaves were eagerly picked from these places.

The flavour of wild Beet, which is the ancestor of all the cultivated beets and spinaches, is rather stronger than most of the garden varieties, and the leaves, being thicker, go further.

BILBERRY   *Vaccinium myrtillus*

Country names: WHORTLEBERRY; HURTS; WHORTS; BLAE-BERRY; BLUEBERRY; HEATHERBERRY.

The fruits of this low shrub have far more local and popular country names than I have quoted above. Bilberries grow on lime-free moors and heaths, and sometimes on the edges of sparse woodland.

The berries are blue-bloomed when ripe and follow the pink, urn-shaped flowers. They are ready for picking in late summer and can be gathered, with the accompaniment of an aching back, until the frosts start.

They are excellent for pie-making and can also be eaten raw, specially with Devonshire, or a thick, clotted cream.

Herbalists recommend them to be used fresh, or dry to provide a natural astringent for those who have stubborn forms of diarrhoea.

Bearberries (*Arctostaphylos uva-ursi*) are the fruits of a related moorland undershrub that grows on high ground. It is far less common than Bilberry and the berries are shiny and red. The leaves of Bearberry are a traditional and effective herbal remedy, when infused in boiling water to make a tea or *tisane*, to help alleviate the distressing symptoms of inflammation of the urinary system.

BISTORT   *Polygonum bistorta*

Country names: EASTER LEDGES; EASTER GIANTS; MAY GIANTS; EASTER MAN GIANTS; POOR MAN'S CABBAGE; SNAKEWEED.

It appears that the leaves of this plant have always been used as a spring tonic and were appreciated in 'green puddings' long before the essential anti-scorbutic properties of vitamin C were understood.

Bistort is a handsome plant and its leaves appear around Easter-tide in some damp meadows, particularly in the north, where its two-feet-high spikes of pink flowers make a high-level carpet in June.

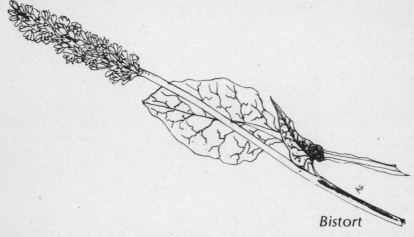

Bistort

In parts of the Lake District, where Bistort leaves are still picked and called 'Easter Man Giants' (a corruption of 'Easter *mangeants*', or 'eating', according to Sir Edward Salisbury), they are made into a spring pudding 'to clear the blood', together with young Nettle tops, a few Dandelion leaves and Wild Garlic.

Green puddings, or dishfuls of well-drained wild plant leaves, need very little cooking and were formerly served compressed into basins, sometimes with hard-boiled eggs halved and arranged on top.

Infusions of Bistort, with its usual accompaniments of Nettles and Dandelion leaves, were made in most country households until even remote districts became too sophisticated. Country children were made to drink this Easter Beer regularly before breakfast.

In Sussex, where Bistort is now a rare wild plant, an old countryman told me that his mother would not let him out to roll his Easter eggs 'down t'hill, until I had downed a half-pint of the nasty stuff'.

It is now known that Bistort leaves are rich in vitamin C and contain vitamin A and some of the B-complex as well.

BITTERCRESSES

Cardamine hirsuta
and
Cardamine flexuosa

Country names: POP-CRESSES; WINTERCRESS.

The unreliability of knowing only local country names is clearly shown by those above. There are other Pop-cresses and at least one other Wintercress (see page 77) so it is important, if you are interested in plants or in using them for free food and health-increasing herbs, to learn their Latin or scientific names too.

It was Gertrude Jekyll, the famous gardener and garden-designer, who introduced me to Bittercresses. She gave a graphic description, and a drawing of one, in her book *Children and Gardens,* which was published in 1908 and which I bought second-hand, thereby getting the clue.

Bittercress
(from above)

Now I use them all through the winter, whole as sandwich fillers for tea, or chopped in salads to augment the short supply of green ingredients. I pull them up whole, cut off the roots and wash the green rosettes thoroughly. The smaller **Hairy Bittercress** (*Cardamine hirsuta*) is not common in our garden, but the lusher **Wavy Bittercress** (*Cardamine flexuosa*) often covers unoccupied surfaces of flower-beds before they are hoed.

The Bittercresses taste like mild Watercress and because of their explosive fruit-pods and their ability to scatter seed, keep on coming up all over the garden.

BLACKBERRIES   *Rubus fruticosus*

Country names: BLEGS; BRAMBLES; BUMBLE-KITES; DOCTOR'S MEDICINE.

Everyone knows how delicious these plentiful wild fruits are and what a pleasure, in spite of the thorns, it is to gather them.

They are more sought after than ever since deep-freezers came into common use, for they come out of the freezer in mid-winter as tasty as they went into it.

Blackberries should be picked while they are firm and dry, and certainly before Michaelmas when the Devil is said to spit, or even urinate, on them.

They are a useful source of vitamin C.

BLADDER CAMPION   *Silene vulgaris*

Country names: COWBELL; WHITE COCK ROBIN; SNAPPERS; FAT BELLIES; BILLY BUSTERS.

The young green shoots of this plant are edible and although it would seem a pity to pick many now that most wild flowers are in short supply, a few would flavour and 'green up' sauces, and some savouries and soups. They should be picked when they are three to four inches high.

Bladder Campion

It is not worth trying the other wild Campions for there are less innocent members of the same botanical family.

In America, a wild Pink (*Silene virginica*) was considered to be poisonous by the Indians, or at best, to be a violent worm-expeller by an American herbalist, which probably means a strong cathartic too.

Bladder Campion is easy to distinguish in flower by its fragile, green, inflated calyces.

BLADDERWRACK *Fucus vesiculosus*

English names: FLOATERS; POPWEED.

This brown seaweed is extremely common. It is edible and can be gathered from shingle, or groynes and rocks between tide-marks. It is also very abundant near the shore on the sea floor.

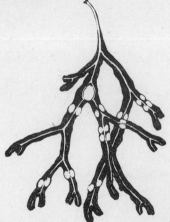

Bladderwrack

Popweed is almost too well known to have to be described. Its strong leathery irregular fronds, with a prominent mid-rib between the pairs of air-bladders which, of course, keep the plant upright even under water, are often picked up on the beach. The stranded seaweeds are usually those that are coarse and have been torn from their roots (known as holdfasts) by rough seas, and it is better to pick fresh Bladderwrack, if you can find a place where it grows.

I have seen sheep browsing on the young tender growth of this seaweed in South Wales at low-tide and their example should be followed by human pickers! It is more pleasant to eat when it is young and the gelatinous fronds should be gently simmered until they are soft.

Bladderwrack tea, made by infusing 1oz (25g) of this brown seaweed in 1pt (550ml) of boiling water, acts as a diuretic and is therefore, taken with care, a help to slimmers.

Most of the edible seaweeds contain vitamin A and some of the vitamin B-complex, as well as beneficial trace elements like iodine that cannot be obtained easily from other accessible plants.

BLEWITS *Tricholoma saevulum*
and and
WOOD BLEWITS *Tricholoma nudum*

Country name: Both are popularly known as BLUE LEGS.

These two edible fungi, which appear in late autumn, both have pale violet-blue stems which makes them easy to identify. Blewits have biscuit-coloured cap tops and gills and grow in grassy pastures, sometimes in rings, and Wood Blewits, especially when young, are more generally lilac, or a smoky violet, all over.

Blewit

Both are good to eat. The latter used to be bought by dyers in Berwick-on-Tweed to provide a source for a blue dye. Wood Blewits are frequently found growing well on dung heaps or compost heaps in gardens, as well as in woods, as their name suggests.

BOG MYRTLE *Myrica gale*

Country names: SWEET GALE; FLEAWOOD.

There is often a clue or two among a wild plant's country names as to its old uses, but the only one in this case is 'Fleawood' which probably means that some parts of this evergreen and aromatic shrub were used in the old days to inhibit the presence of the all-too-common little blood-suckers from persons, or from their bedding, or even their animals. I cannot find any more exact hints.

Bog Myrtle bushes grow plentifully in some wet heaths like, for example, those that remain in the New Forest, and tradition has it that its leaves were used for flavouring ales before:

Hops, reformation, bays and beer
Came into England all in one year.

The year was 1524 and many have held since that better ale was made before Hops were brought into this country.

Bog Myrtle ale seems to have been made since Anglo-Saxon times and some enthusiasts still make their own. If you have a plentiful and easily available supply of this plant it is worth trying. You need enough of the small tough leaves almost to fill a crock. Pour water over them till they are covered, then strain it into a pan and add half a pound of honey to each gallon. Bring to the boil and then pour this sweetened water back over the leaves, pushing them well down with a wooden spoon or old-fashioned wooden potato-masher.

Add more boiling water to top up, so that the leaves are covered, but without any more honey, and let it all stand until it is only warm. Stir in yeast gently until the water starts frothing. Let all this stand and 'work' for six to eight days, then de-froth it and strain it off into a cask. *Do not cork the cask,* until the ale has stopped 'working', and *do not try any bottling* unless you have strong stone jars. It is not fit to drink until it has stood for at least a month, or more if the weather is warm.

BORAGE   *Borago officinalis*

Country names: BEEBREAD; VEGETABLE ICE; BLUE STARS; COOL TANKARD; OX-TONGUE.

This beautiful cobalt and starry-flowered herb was said to have been brought to Britain by the Romans. It flourishes in many gardens and never needs re-planting although it is an annual, for it seeds itself. It is also often found growing freely, where it must have once been chucked out, on wasteground and on rubbish-tips and occasionally in hedgerows.

Borage has the reputation of being a cheering or cordial herb. Pliny said that 'I, Borage, bring alwaies courage' and modern herbalists believe that this is true because the plant acts on the adrenal glands and stimulates them.

Borage

The young leaves taste a little like those of cucumber and can be chopped raw for sandwiches or salads, or be cooked with cabbage to improve its flavour. The flowers can be crystallized or candied (see page 117), or used fresh in salads or to float in Pimm's No. 1, claret cup, white wine and even homemade lemonade.

Borage tea or *tisane* is said to make a soothing eye-lotion or when drunk, to provide a mild laxative and liver stimulant as well as to increase a nursing mother's milk.

BROOKLIME   *Veronica beccabunga*

Country names: BIRDS-EYE; HORSE WATERCRESS; BITTERCRESS.

At one stage in our history many of the plants that resembled or grew in similar places to others that were frequently used, like Watercress and this Horse Watercress, were given the epithet 'Horse'. (See also Horseradish and Horse Mushroom.) Etymologists do not think that horses necessarily had to be fond of these, but that the description may merely mean that these plants could also be bigger or coarser, or even locally commoner.

Certainly most of the blue-flowered speedwells, of which Brooklime is one, have earned the name of Birds-eye at one time or other, although the curious thing is that very few birds have blue eyes, the jay being a notable exception!

However, Brooklime Speedwell has been known as a blood-purifying herb for centuries and Gerard seems to have known about it as an anti-scorbutic. 'Take the juice of Brook-lime, Watercresses and Scurvy-grass, each half a pint; of the juice of Oranges, four ounces; fine sugar, 2 lbs, make a syrup over a gentle fire: Take one spoonful in your Beer every time you drink', he said, 'to cure the Scurvy'.

In comparison with Watercress, with which it often grows, Brooklime is a little bitter. It seems, however, to be commoner in some places than Watercress and to grow in damp, less polluted places. It is worth trying as a salad plant.

BROOM   *Sarothamnus scoparius*

Country names: BESOM; LADY'S SLIPPERS; GENISTA.

Broom is a tall, pliant and rather shapeless shrub which bursts into golden pea-shaped flowers in the countryside, and on the sides of cuttings along roads and motorways, in May. It has many links with British history.

For instance, the Plantagenets took their name from it and so did many places, such as Broomershill, Bromsgrove and Bromley and others, where it must have grown abundantly. It was credited with magic powers and witches rode brooms made from this plant.

Remember this if you eat it and treat it accordingly, for eaten to excess Broom can be harmful! But a few Broom flower-buds can safely be added to salads, or they can be pickled.

Medicinally, young broom branch-tips have frequently been used by herbalists to make a diuretic drink and Gerard spread the gossip of previous times by writing about this: 'That woorthie Prince of famous memorie

Henrie the eight King of England was woont to drinke the distilled water of Broome flowers against surfets and diseases thereof arising'.

Broom, of course, was well-named and a few years ago I lived near an old Sussex farm-worker who walked miles to pick some Broom branches every year, 'To make me old gal a new Besom, see!'

BULLACES   *Prunus domestica*

Country names: WILD PLUMS; CHRISTMAS; CRYSTALS; KESLINGS

The fruits of this wild plum tree ripen very late in the autumn and are always sour. But they have always also been eagerly sought for making 'Bullace Cheese' or 'Bullace and Quince Jelly'.

They can be found in hedgerows and on the edges of woods and the trees still grow in some old-fashioned gardens.

Bullace fruits make a good wine which, if kept long enough, is said to be like Port. For hundreds of years they have also been made into country electuaries and used for syrup, to be put by in case any member of the family should be in need of a 'gentle purge'.

CHERRY PLUMS   *Prunus cerasifera*

This species of wild plum is less common than Bullace trees, but where it does grow, it is often in long hedgerows which all come into white flower together early in spring. The fruits are earlier too and are round and golden-yellow when ripe and ready for picking.

Both this plum and the Bullace were the original parents of our large, fleshy-fruited garden plums.

BURDOCK   *Arctium lappa*

Country names: SWEETHEARTS; BURRS; DONKEY BURRS; BACHELORS' BUTTONS; BEGGARS' BUTTONS; CUCKOOS' BUTTONS; TUZZY-MUZZY; WILD RHUBARB.

Burdock leaves are very like Rhubarb leaves in shape, but that is all, for when they first appear early in the year, they are hairy and a pale ice-green which can almost look blue. They get greener as they grow and spread out, but they never lose their downy surface entirely.

The young leaves and steams are both edible. They need washing thoroughly before they are lightly boiled. I use a steamer for these and other soft leaves, as I prefer to keep them out of water in case their virtues are unnecessarily diminished.

It may be of interest to know the different 'virtues' of Burdock, for a modern nutritionist finds that this plant contains vitamins A, B and C as well as traces of calcium, phosphorus and iron.

The way, of course, to get the maximum benefit from tender Burdock growth is to chop it and eat it raw in salads, or to shred it and juice it in a blender, either alone or with other valuable wild plants.

Burdock

Burdock, with its globular inch-wide hooked fruits which can be such a pest when they attach themselves to walkers' clothes or to sheep and shaggy dogs' coats, is a common plant in most parts of the country. Actually, the fruits can be collected and kept dry to be used during the winter as a supply, with other wild seeds like Mustard, Alfalfa, Carrot, Dandelion, and Garlic, to germinate and sprout and eat whole (see Alfalfa, page 17). But Burdock seeds must be removed from their hooked outer fruit-cases before use.

BUTCHER'S BROOM   *Ruscus aculeatus*

Country names: KNEE HOLLY; BOX HOLLY; JEW'S MYRTLE; SHEPHERD'S MYRTLE.

This curious dark and spiny evergreen shrub with its tough, short, flat leaf-like branches can be found growing abundantly in some woods and hedge-edges. It is what is known as a 'locally common' plant and is well worth studying closely when you come across it.

As an edible plant it should not be gathered unless it is in really good supply for there are many areas where it is now considered to be rare, so it has to be conserved where it does grow.

Butcher's Broom

Presuming, however, that it is very common where you live, it is worth picking one young (and bright green shoot) in the spring, when it is not more than five inches tall, and cooking it like Asparagus to see how you like it. But if it does please you, always remember that the plant cannot go on thriving without some young shoots growing on to replace the old woody ones.

Sometimes the small greenish-white flowers in the centres of the old flat branches set into big red solitary berries. They are attractive but BEWARE of them for they are extremely purging and should *not* be eaten.

Butcher's Broom used to be used, it is said, for scouring butcher's tables but the old countryman I knew used to pick a bunch whenever he wanted to clean his chimney.

He tied the prickly bunch to the top of a long pole and then 'pushed it up the chimbley, to get it nice and clean-like'.

CARRAGEEN   *Chondrus crispus*

Common names: IRISH MOSS; SLOKE (see also Laver).

An edible red seaweed that grows in the intertidal zone all round our coasts, Carrageen is used industrially, like the Kelps (see page 74), in providing agar or vegetable gelatines for a great number of purposes.

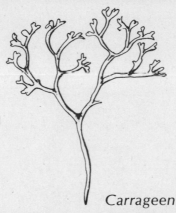

Carrageen

It is best gathered in the spring and should be gently boiled until reduced to a jelly which can then be used as a nutritious addition to many foods. It can be dried, when freshly picked, and kept like that, to be used at any time.

CARROT   *Daucus carota*

Country names: WILD CARROT; BIRD'S-NEST; PIG's PARSLEY.

Wild Carrots' flowers are easily recognized by their red central floret which shows up among all the surrounding lace-like white ones.

The root is tough and although it is the ancestor of modern, juicier kinds, it is not worth gathering.

Wild Carrot leaves make a good *tisane* with a tonic and blood-purifying, antiseptic and liver-stirring effect on sluggish systems, and in the late summer if a few fruiting heads (which do resemble birds' nest when they are brown and curled inwards on the plant) are collected, the seeds can be shaken out of them when they are quite dry and used during the winter as valuable germinating sprouters. (See Alfalfa, page 17, for instructions on sprouting.)

CEP   *Boletus edulis*

Country names: STICKY-BUN FUNGUS; Cépe (France).

Ceps grow in woods and look like stalked sticky, brown-topped buns. Instead of having gills under their rounded tops, like mushrooms, they have a finely-tubed white or pale yellow spongy surface.

Cep

There are other species of *Boleti* which are also edible – indeed, most of them are – but they need careful study as some of them might upset some people, and these should obviously be avoided by them. Fresh Ceps contain vitamin D.

CAULIFLOWER TOADSTOOL   *Sparassis crispa*

Country name: BROODY HEN (Switzerland).

At first sight anyone could be forgiven for thinking that this intricately curled, edible fungus was a tightly crumpled sheet of yellowing paper, but a closer look reveals that it is growing out of an old pine stump; certainly it does bear a resemblance to the top of an elderly cauliflower which has lost its crispness.

Young Cauliflower Toadstools should be cut carefully

from their bases, rinsed in many waters, separated and fried in batter, or casseroled in pieces.

CELERY   *Apium graveolens*

Country name: SMALLAGE.

The typical strong green colour, shape and smell of the leaves of wild Celery is easy to recognize. It is a tough plant which grows near the sea, and if you are guided by the fact that grazing cattle leave it alone, it may not appeal to you. But remember that it may be too flavoursome for them, specially while there is plenty of young grass about.

BEWARE of confusion with **Hemlock Water Dropwort** (see page 16), a poisonous and not very different-looking Umbellifer, and be careful of your identification. The smell of Celery from broken stems and bruised leaves ought to be enough for anyone who is conversant with cultivated Celery.

Young wild Celery leaves are delicious chopped in salads or sandwiches, and some people also cook them in sauces, soups and casseroles.

Another species in the same tribe, **Fools' Watercress** (*Apium nodiflorum*), sometimes called Cow Cress, grows in fresh water, often with Watercress. The leaves of this plant are also edible and make a good flavouring herb despite frequent stern warnings, which are also applied to the bigger wild Celery, to the contrary.

The Celeries, medicinally, are slimming herbs because they 'provoke the urine' or, in other words, are diuretic. They also have tonic, digestive and antacid virtues.

CHAMOMILE   *Anthemis nobilis*

English Chamomile only just creeps into the category of 'free wild plants' or weeds because, like a few other plants I have included, it grows on strongly and spreads if it is thrown out of gardens. It also spreads out and away from the edges of a Chamomile lawn or from a herb bed if it is given the chance.

When you are certain of this plant's identity, collect and dry the flower-heads for one of the most useful of all herbal teas.

Chamomile tea, or *tisane*, sipped hot at bed-time, with the addition of a little honey if you crave for sweetness, will give a good night's sleep as well as a calming drink for nervous, stressful systems in the day-time. It is a digestive herb and safe enough to give to babies when they are fretful or teething.

Chamomile

Culpeper said that 'a decoction made of Chamomile and drunk, taketh away all pains and stitches in the sides'. He also extolled its soothing possibilities for the bath – if you have enough, try boiling up a double handful of this herb for a few minutes and then straining the resulting liquid into a hot bath.

A Chamomile rinse is said to improve fair hair and 'make it glint like gold'.

An old Kentish woman, whom I knew when I was a child, used a poultice made with Chamomile flowers as a cure for ear-ache: 'Lay your poor ear on it, dearie, as hot as you can bear!'

Chamomile lawns are rarer now than they used to be when most large gardens had their own in the leisured days. Sir Francis Drake was said to be playing bowls on one just as the coming of the Armada was sighted.

A few fresh green leaves of Chamomile add a 'different' flavour to Cabbage if they are chopped sparsely with it while cooking.

CHANTERELLE   *Cantharellus cibarium*

Chanterelles are almost unmistakeable edible fungi that are more or less horn-shaped, though they sometimes have irregularly lobed and uneven tops. They are a deep apricot yellow all over, including their gills.

These fungi smell of apricots too, if their flesh is handled or broken, and they grow in woods from July until November, or even in some mild years, December, as they seem able to tolerate some degree of frost.

Chanterelle

This is one of the most sought after and famous of the esculent fungi and only confusable with False Chanterelles which are the same colour, harmless, and grow in coniferous woods. These do not smell of apricots at all and although also edible are most disappointing, tough and tasteless in comparison with true Chanterelles, but are thought to contain, like the others, vitamins A and D.

CHARLOCK   *Sinapsis arvensis*

Country names: WILD MUSTARD; RUNCH; CALVES'-FEET.

Everyone must know this crude yellow-flowered weed. It comes up at the edges of fields, even among grass, in waste places, by the roadside and on rubbish-tips, often accompanied by a scattering of Red Poppies.

Red Poppies, incidentally, if common, will provide ripe seed for decorating home-made bread crusts, or for infusing (one teaspoonful to a pint of boiling water) to make a sleep-inducing drink.

Charlock leaves are edible if cooked like Spinach but can be too strong and tough. I only collect their ripening fruit-pods and save the seeds for sowing with other sprouters (see Alfalfa, page 17) in the winter.

There are old wives' tales which may for once be true, which say that all members of the Cabbage, or Cruciferous, family which have their four petals arranged in the shape of a cross are edible. This is by no means a totally safe clue, for other plants in quite different groups may grow with only four petals as well. There are a large number of Crucifers, however, including mustard, the cresses, cabbages and kales, as well as a variety of smaller and less significant weeds, which are all safe to eat.

CHERRIES   *Prunus avium*

Country names: MERRIES; GEANS; GASKINS; MAZZARDS.

As the fruits of the wild Cherry, however sour they may be, are eagerly sought by wild birds, you will be lucky if you manage to harvest any. However, the beauty of this forest-intruding tree belongs to its flowers in spring, and A. E. Housman was right when he called it: 'Loveliest of trees, the cherry now is hung with bloom along the bough'.

The Romans are credited with introducing cultivated Cherries into Britain and since those days many more varieties have been grown here to improve our fruit-stocks.

Bulges of resinous gum that exude from Cherry tree-trunks used to be dissolved as a cure for coughs and are safe for chewing.

Modern research into the attributes and virtues of plants follow, in this case, traditional ideas that cherries of all kinds, even though tasting acid, are excellent for those who suffer from arthritis and gout, as well as for stomach ulcers, as is the gummy resin on the tree's bark.

The fruits of the **Bird Cherry** (*Prunus padus*) are small and bitter and hardly worth seeking. **Cherry Plums** are very good and plentiful in some hedgerows (see page 32, Bullaces.)

CHERVIL   *Anthriscus sylvestris*

Country names: KEX; ADDER'S MEAT; COW PARSLEY; QUEEN ANNE'S LACE; RABBITS'-FOOD.

Wild Chervil is one of the first plants to 'green up' roadside banks and hedgerows as the delicate clear green leaves start showing from January onwards. These tender leaves are eagerly picked by rabbit owners and others with poultry and with ponies which are unable to get at the fresh green grazing for themselves.

It is the first of the big confusing Umbelliferous plants to start growing in spring and its identity should be checked and learnt from an expert before it is picked to use either raw or cooked.

The smell from the broken stems of many members of this vast family, which does include poisonous members like **Hemlock**, can be helpful but is not a wholly safe guide to the different species. Hemlock, smelling as some say, of mouse-dirt and with dark fern-like leaves and blotched stems, ought to be distinctive enough to avoid, but it is wiser to learn about these plants thoroughly so as to be certain of their exact identities.

The young leaf-bearing stems of wild Chervil are useful as flavourers in salads, soups, casseroles, omelettes, and so on, but there are people who are allergic to its juices, whose hands blister when gathering it. Although this seldom happens, do not use it for a culinary herb if it should ever give you a skin-rash just by touching it.

CHESTNUTS   *Castanea sativa*

Country names: SWEET CHESTNUTS; SPANISH CHESTNUTS; HEDGEHOGS.

The fruits of this tree are ripe in October in the south of England and soon begin to fall. There are gathered by squirrels, mice, deer, rooks, and jays as well as by humans.

It is worth taking a pair of old and thick gloves if you go out especially to collect them, as breaking open their spiny outer fruitcases can sometimes be difficult even after they have been rolled underfoot.

The nuts have been used in many countries for making flour, or meal, which was then made into puddings, cakes and bread. Nowadays it is probably seldom used in this country for much more than making stuffiing for Christmas turkeys, although gorgeous *Marrons glacés,* in which whole, peeled chestnuts are candied, drained and dried again and again after boiling in a strong sugar solution, are well worth the effort of making.

In America, where the native edible Sweet Chestnut has almost completely disappeared from the New England scene, introduced species from China and Japan are now doing well and will soon replace the necessity for importing these nuts from Europe.

Sweet Chestnuts are crisp and flavoursome when eaten raw or roasted, and boiled Chestnuts are one of the pleasantest accompaniments for Guy Fawkes, and Hallowe'en parties. They are nourishing, even if starchy, and have a good mineral content and are low in fats.

CHICKWEED   *Stellaria media*

Country names: CHICKENWEED; WHITE BIRD'S-EYE.

Some gardeners use young weeds, before they start seeding, as green manure and dig them in. Others, at the same stage, use them to augment the compost heap, but on the whole most useful plants like Chickweed are thrown out, or burnt or destroyed, which is wasteful.

Fresh young growths of Chickweed make a good vegetable. I strip the leaves off the main stems and cook them in salted, boiling water for two minutes. I also use tender plants raw in salads, or as sandwich fillers.

Caged birds and shut-in poultry both benefit from Chickweed and wild birds seek it out, particularly for its seeds.

Culpeper suggested that it was good for all kinds of inflammations and it is a soothing, demulcent herb from which a green ointment can be made to anoint rheumaticky joints and chilblains (see also Elder ointment, page 55, and Primrose ointment, page 96). It is easy to make by simmering $\frac{1}{2}$lb (225g) of washed Chickweed in 1lb (450g) of pure lard, until the lard starts turning green. Then strain the leaves out, let it cool a little and pot up.

Old-fashioned pig-keepers used to give their runts or

weakly young piglets with diarrhoea, or 'the scours', Chickweed added to their mash to provide them with natural iron.

CHICORY   *Chichorium intybus*

Country names: SUCCORY; BLUE SAILORS; OUR LADY'S EYES; GRETCHEN-WEED.

Wild Chicory, with its inch and a half, day-lasting blue flowers is usually an escape, or merely a naturalized plant in Britain. It is less common than it was now that roadside verges, where it often used to grow, are frequently kept cut down and over-tidy.

Chicory

It therefore seems a sin to pick more than a few leaves to try as an additional flavour in salads or to attempt to dig up roots to blanch, in pots, for their shoots or 'chicons' to grow as a winter salad plant.

The roots of cultivated Chicory are dried and ground to make popular coffee substitutes and all parts of the plant are slimming to eat because of their diuretic properties.

CLEAVERS   *Galium aparine*

Country names: CLIVERS; CATCH-GRASS; LITTLE SWEETHEARTS; GOOSE-GRASS.

Like Burdock, Cleavers has hooked fruits which give the plant the old name of Little Sweethearts, as they too cleave to all human and animal passers-by. But Cleavers' leaves and stems are hooked too and country children sometimes throw a handful at each other, or at an adults' back, hoping that they may stick on and not be noticed.

Cleavers is a common annual herb with weak stems which scrambles aloft by clinging to other herbage in the hedgerows. It comes up early in the year and is a useful green vegetable if picked young and boiled only very quickly in salted water. Like all wayside weeds, it needs careful washing.

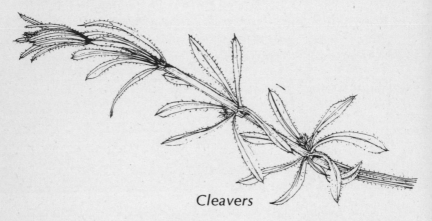

Cleavers

Cleavers has the reputation of making 'they who are stout grow lean' and can be taken as a mild diuretic tea if infused, after measuring it out as a good pinch of fresh or dried herb between the thumb and forefinger, with ½pt (275ml) of boiling water poured over it. It tastes very refreshing and has a slight flavour of cucumber.

'Gross women', according to an old herbalist, should 'Make a pottage of cleavers with a little mutton and oatmeal, to cause lanknesse and keep them from fatness'.

The ripe seeds, though small to pick, can be ground and used as a substitute for coffee (see also Acorns and Chicory, pages 15 and 43).

CLARY   *Salvia horminoides*

Country names: WILD SAGE; CLEAR-EYE; EYESEED.

Clary grows on banks, hillsides and sloping pastures where the soil is well-drained. It can often be seen on steep railway or road-cuttings through chalk hills.

The purple flowers and bracts, smaller than those of garden Sage, can be used to decorate and flavour salads and the young leaves make good flavourers. An infusion of the young leaves is sometimes used as a gargle or even a digestive tea.

The seeds of this plant were once used to help to strengthen dim eyesight. They were put, one at a time, into the eye of anyone complaining that a foreign body had got into their eyes, because they were reputed to 'absorb the humours that spread over the eyes'. Actually the seeds absorb moisture and swell and might thereby have taken a bit of dust along with them before they were carefully removed. Hence their old name 'Clear-eye, and the subsequently derived English name of Clary.

COMFREY   *Symphytum officinale*

Country names: KNITBONE; CHURCH BELLS.

Comfrey has a wide variety of helpful uses. To start with its medicinal virtues; the big leaves can be made into a green, demulcent poultice, or an ointment or even a lotion (see page 90), to speed up the re-knitting of fractured bones that are not in plaster of Paris. Poultices from these leaves, or indeed the ointment or lotion, can be used to help in reducing swellings, strains and sprains and bruises.

The young leaves can be steeped in boiling water to make a Comfrey tea which is helpful when drunk by those 'with coughs and phlegm on the chest'. Modern herbalists and naturopaths find that this herb is as much of a 'heal-all' as their older forebears claimed and that Comfrey tea does help those with respiratory troubles, as well as anyone with gastric ulcers.

This plant is the only known source of *allantoin*, which is a quite remarkable healing agent.

Young Comfrey leaves 'frittered' in batter, make a pleasant type of pancake, and the leaves can also be boiled as a plain green vegetable.

Comfrey is a common wild flower of damp fields, hedge and ditchsides. It grows up to three feet tall and has creamy, pinkish or dingy-purple, half-inch long, thin bell-shaped flowers which grow in a row in a crook-shaped cluster. This flower-crook opens up gradually so that the first flowers at the bottom fall off as those tighter in the curl, expand and open.

Comfrey

Occasionally colonies of an even taller, bright stained-glass-coloured, at first pink and then bright blue-flowered, Russian Comfrey are found. These are relics of crops of Russian Comfrey that were until recently frequently cultivated in Britain as a forage crop or for green manure. Unfortunately the practice has diminished in these days of farming with prepared and concentrated fertilizers.

CORNSALAD   *Valerianella locusta*

Country names: LAMBS' LETTUCE; RABBITS' LETTUCE.

There is nothing about this small herbaceous plant that resembles Lettuce, except perhaps its leaf-colour and the texture of the long, oblong leaves.

It grows in fairly dense colonies, even in the wild, on dry banks including railways banks, and is always worth picking for salads. It can be grown in gardens as a winter crop of tender greenstuff. The seeds may initially be purchased from herb-growing nurserymen, and saved each year for re-planting subsequently.

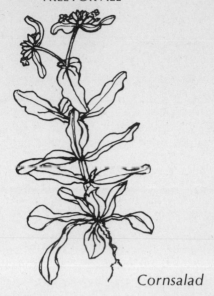

Cornsalad

CORN SOW-THISTLE   *Sonchus arvensis*

Country names: DOG THISTLE; MILKWORT; RABBITS' VITTLES; SWINE THISTLE.

When the wide bright yellow shaggy flowers of this corn weed first come out, Corn Sow-Thistle is a handsome plant. It looks more ragged when the blooms fade and are replaced by cottony white fruits. The leaves should be picked for eating while the plant is still young.

The Sow-thistles have milky sap which shows when the leaves and stems are broken. BEWARE of all the poisonous **Spurges** with equally milky juices that are sometimes called Wartworts, and used to be used to dab on warts. Their juice is corrosive and can burn delicate tissue, especially if it gets into the eyes.

COWBERRY   *Vaccinium vitis-idaea*

The fruits of this northern wild moorland plant, which is related to the Cranberry, are red and very sharp in taste, but are high in vitamin C content and are used for jelly-making. Crab-apples, or windfall garden apples are added, plus, of course, 1lb (450g) sugar to each pint (550ml) of fruit pulp.

COWSLIP   *Primula veris*

Country names: ST PETER'S KEYS; PAIGLE; CULVERKEYS; MARY'S TEARS; PALSYWORT.

Now that natural grassland and downland meadows have been ploughed up to make hedgeless acres for agricultural crops, Cowslips are becoming scarce. It is beginning to be difficult to remember how golden Cowslip heads used to cover the ground, so that a child could say:

> Thus I set my printless feet,
> O'er the Cowslips' velvet head
> That bends not as I tread.

In those days, of course, there was no harm in picking the fragrant flowers for wine-making, or for using in salads, or even for pleasure, but now that there are so few about, except in a few extremely remote places, such occupations as picking basketfuls of the lovely spring wild flowers are finished.

Cowslip

If you are fortunate enough to live in an area where there are still some Cowslips, try to instigate a campaign to conserve those that are left so that most of their flowers can set and produce seed for future young plants. A few could be picked perhaps and their flowers dried to save for making sleep-inducing, soothing herbal tea. This old folk-remedy, 'Cowslip tea', had the reputation for helping those with vertigo, or anyone who suffered from buzzing in the ears.

CRAB APPLE　　　　　　　　　*Malus sylvestris*

Country names: WILDLING-TREE; SCARB; SCRAB; SOUR GRABS; BITTERSGALL; CRABS.

Some of the old country names for Crab Apple trees are picturesque in that they conjure up an image of the screwed-up faces that were caused by taking a bite out of a freshly-picked fruit.

'Crabs' are, though, edible and can be used to provide pectin to set other wild fruits for jam, or jelly-making, as well as by themselves for Crab Apple jelly, or for wine. I always think they need added flavour in the way of Elder-blossom, or even lemon-juice, but when their jellies are cooked and cleared they set excellently.

True Crab Apples are small and green. They are round but flatter at the top and bottom. Some trees may produce bigger, coloured fruit and these are probably seedlings from nearby orchard apples, especially, as Alan Mitchell says in his book, *A Field Guide to the Trees of Britain and Northern Europe,* 'if the flowers are very pink'.

In the old days Crab Apple juice was known as *verjuice,* an ingredient that appeared in cookery books even until Victorian times.

CROWBERRY　　　　　　　　*Empetrum nigrum*

The black ripe fruits of this heathland plant, which grows particularly on acid/moor areas of land, are aptly named, according to one writer, for being, apart from their vitamin C content, better for 'the crows, than humans'!

DANDELION　　　　　　　　*Taraxacum officinale*

Country names: WET-A-BED; SHEPHERD'S CLOCK; TIME-TELLER; MONK'S-HEAD.

At least Dandelions are still common and so far no conservationists frown when they see people picking the golden miniature 'suns' for wine-making in the late spring.

But I would not even now waste a Dandelion if I had a good root in my garden. Potted up, in the winter, and kept in a dry warm place, in the dark, a Dandelion or two can keep a small family in young, blanched, tender salad leaves.

Dandelion

'Dandelion coffee', made from dried and ground Dandelion roots, makes a good drink which is free from tannin, relaxing and splendid for 'They who be liverish'. This plant has also mild diuretic and laxative properties, so some of its graphic country names above, of which there are so many that those included are a mere sprinkling, could easily have been found to be true if anyone had eaten too many.

Dandelion leaves are rich in vitamins A and C and also contain vitamin B1.

DEAD-NETTLES   *Lamium album*
                                and
                        Lamium purpureum

Country names: *For White Dead-nettles:* BEE-NETTLE; HELMET-FLOWER; SUCKY SUE. *For Red Dead-nettles:* DUMB-NETTLES; RED ARCH-ANGELS; BUMBLEBEE-FLOWER.

The leaves of both these Dead-nettles can, if very abundant, be cooked as a green vegetable.

In America, a few young Dock leaves and Nettles are added to a green 'health beer', made from these and other plants. They are juiced in an electric blender and a curl of orange or lemon peel is put in to increase the flavour.

DEWBERRY   · Rubus caesius

Country names: BLUE BRAMBLE; TOKEN BLACKBERRY; EARLY BLEGS.

Dewberries are ripe before Blackberries, but are seldom as plentiful, although they are delightful to find with their large, fewer, blue-bloomed drupes showing up clearly on their low, thornless bushes.

They can be used in all the same ways as Blackberries but, to me, they never have quite such an excellent flavour.

DOCKS   Rumex crispus
and
Rumex obtusifolius

The leaves of the Curled-leaf Dock and of the Broad-leaved Dock can be cooked as a green vegetable if found in a young and tender state, but are usually too bitter. If desperate for freshly picked greens, pour boiling water over them, and pour it off again, before cooking them in boiling and slightly sugared water for two to four minutes.

Homoeopathic medicines include potentized Rumex crispus to be prescribed for people who regularly seem to have acute runny noses and hay-fever symptoms very early in the morning, on first getting out of bed.

See 'green health-beer' under Dead-nettles above, to which Dock leaves can be sparsely added.

DULSE   Rhodymenia palmata

This tough, edible red seaweed, which grows on many of our coasts, needs careful washing and slow gentle boiling. It can, however, be eaten raw and is good to give children to chew instead of chewing-gum or bubble-gum.

Seaweeds, or sea-vegetables, are exceptionally rich in

minerals and vitamins and are a good source of protein. The Iroquois Indians used to use them to replace salt.

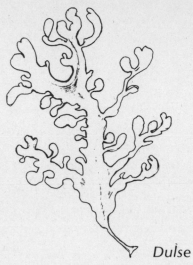

Dulse

Once you are used to the idea of eating seaweeds, their flavours will interest you and you will find many ways of introducing them to your meals to add relish that you hardly knew existed before.

DUNGWEED   *Chenopodium album*

Counyry names: LAMBSQUARTERS; WHITE GOOSEFOOT; WILD SPINACH; ALL-GOOD; PIGWEED; MELDWEED; MUCKWEED; FAT HEN; *Grasse-poulette* (France).

This common farmyard and garden weed has been neglected as a free vegetable since our cultivated forms of Spinach have replaced it, but our ancestors picked and ate it regularly for anti-scorbutic 'greens'.

The young leaves can be eaten raw, but as the plant matures and is often found growing on somewhat dubious dungheaps, they are better lightly boiled.

Dungweed grows abundantly in Britain wherever there are suitable rich conditions for it. Its popularity with wild birds can be seen if ever it is left to seed, when flocks of greenfinches or goldfinches soon find it. If you have a caged bird, it is worth picking a few of these ripe fruiting heads to shake out the seed for it.

Dungweed

BEWARE of eating the roots of this plant and of **Orache**, (see page 91) as they contain saponins which can have a toxic effect.

ELECAMPANE   *Inula helenium*

Country names: ALICAMPANE; SUN-FOLLOWER; WILD SUN-FLOWER.

Elecampane is one of the old herbs that were probably brought to this country by the Crusaders as a help for themselves and for their horses. It was certainly a popular herb in Elizabeth I's time and has since spread out of some gardens into hedgerows and on to stream-sides. It is tall and unmistakeable with its sun-following, yellow, big daisy-like flowers that are heavily surrounded by green bracts.

It is included here as, although usually uncommon, it can flourish well locally where it has long been considered a wild weed.

Elecampane was named, botanists think, in the sixteenth century after the 'lamentable and pitifull tears of Helena, wife to Menelaus, when she was violently taken away by Paris into Phrigia, having this herbe in her hande'.

Elecampane

It has the reputation of being good as a cure for a tiresome cough and Elecampane roots used to be candied, or dried and powdered as an asthma cure. It was also used 'to keep serpents and evil spirits away'.

ELDER   *Sambucus nigra*
Country names: DEVIL'S WOOD; DOG-TREE; GOD'S STINKING TREE; JUDAS-TREE.

The Elder, which everyone knows and some foolishly scorn, has earned itself a wide variety of popular names in various parts of the country. Only a few are mentioned above.

Most real country people appreciate this tree even if it does grow too abundantly and often assumes ugly, lolling or gaunt positions. They know that delicious white, sparkling wines can be made from its fragrant creamy panicles of flowers, which can also be broken up and dipped in batter to fry as elder-flower fritters.

Elder-flower tea, made by pouring boiling water on to these blooms, is refreshing and some people added honey and yeast to make a longer-lasting and sparkling summer drink. Elder blossom adds a good flavour to stewed rhubarb and gooseberries and although its season is short, the flowers can be dried for later use.

Elder-berries are well worth picking for making red winter wine as well as to improve the endless and inescapably dull taste of stewed apples. I put several bagfuls in the freezer this past year and have been using them to add to other fruit ever since.

Elder

The berries, boiled together with a few drops of peppermint, make a useful cordial, with a little honey added, for helping to get rid of persistent colds and coughs.

In the spring the young leaves can be gathered to be boiled in pure lard as a healing ointment for external inflammations, as well as for 'cows' sore bags'. Enough leaves should be added to the boiling fat to make the liquid 'as green as grass', when it should be strained, allowed to cool and then potted up in jars with air-tight lids.

EVERGREEN ALKANET   *Pentaglottis sempervirens*

Evergreen Alkanet grows well in some hedgerows, and on many waste-places and rubbish-tips, having originally been thrown out of gardens. It is a harsh, hairy, long green-leaved plant with small cobalt-blue flowers which are frequently confused with the far bigger flowers of Borage.

These flowers are harmless and may be used for decorative purposes in salads or drinks.

FENNEL   *Foeniculum vulgare*

Country names: FEATHER-FENNEL; FISH-FLOWER.

Fennel grows wild near the sea, especially on cliff-tops or along roads leading to the shore. It has fine, almost hair-like leaves, and umbels of minute yellow flowers. The whole plant is noticeable because of its frail, though tall, appearance.

Fennel is said to have been popular with the Romans who chewed its stalks (to cure bad breath, as it is supposed to do?) and also crowned their heroes with crowns woven from it.

Fennel seeds as they ripen are, in fact, delicious to nibble. They were one of the 'go-to-meeting-seeds' like those of Angelica or Sweet Cicely, used to alleviate boredom during long religious meetings in the eighteenth century, and ought still to be carried in a little snack box by those who have a perpetual urge to eat sweets, or biscuits or cakes. These seeds are non-fattening and also have a reputation for curing depression! They were used to flavour gripe water for babies.

Both Green and Bronze Fennel grow well in gardens and a few snippets of young stems can be boiled as 'poor man's Sparragrass', or used chopped raw in salads. The leaves are used for flavouring sauces; indeed, judging by the way that cooks extolled their virtues in the old days, they must have been used to disguise tasteless or even 'high' fish.

Witches and ghosties both fled before the presence of Fennel Seeds when they were made into stiff pastes to block up cracks in the door, or key-holes, 'against their entering'. As this plant was also used to 'restore good temper' it was, or ought to have been, frequently sought, especially after long, cold winters when as Culpeper promised 'It absorbeth all the flegmatic humours'.

FEVERFEW   *Tanacetum parthenium*

Country names: BACHELOR'S BUTTONS; DEVIL DAISY; FEATHERFEW; NOSEBLEED; HEADACHE-PLANT.

This plant's English common name points out its old use as a fever-reducer or febrifuge. It was also known as the vegetable aspirin and even some modern herbalists maintain that Feverfew Tea can alleviate headaches, including some types of migraine. They also say that it helps those with high blood pressure.

It is therefore worth picking and drying to keep as a *tisane*-making standby.

Feverfew

Feverfew comes up, unsown, in many country gardens where its Chrysanthemum-like but lime-green leaves, and spreads of little white, single daisy flowers are an asset to unregimented borders and beds, even if its presence is thought, by those who are knowledgeable, to indicate the presence of a poor soil!

FIREWEED   *Epilobium angustifolium*

Country names: ROSEBAY WILLOW-HERB; FRENCH WILLOW; BLOOMING SALLY; BOMBSITE-WEED.

Great colonies of this magenta-flowered wild plant often come up to cover recently cleared areas in woods, or places where there have been ground fires. Hence, of course, its common name.

It romped up on waste places that had been bomb-cleared in cities during and after the Second World War and helped to cheer everyone up by the vivid colour of its flower-spikes.

The young leaves, as long as you are sure of their identity before the plant flowers, can be picked and cooked lightly as a green vegetable.

GARLIC   *Allium sativum*

Country names: RAMSONS; STINKERS; BADGERS'-FLOWER; GYPSY'S ONIONS.

Wild Garlic is unmistakeable once it has been trodden on or bruised so that its smell escapes into the air! In the distance the twin, bright green, Lily-of-the-valley-like leaves might indeed be those of the deliciously fragrant plant, but the illusion quickly vanishes!

Lily-of-the-valley is poisonous, so that is a slight recompense, as wild Garlic, when the leaves are young and tender, is excellent chopped up raw for adding to sandwiches, coleslaws, green salads and as a dressing to sprinkle over cooked vegetables. It is also a good flavouring for sauces, soups, casseroles and roast meat, and enhances their own flavour – for anyone, that is, who is a Garlic addict.

Wild Garlic has many medicinal properties and is thought to be good for high blood pressure, for digestive troubles, for 'those with noisy breathing' as well as for 'them who would be as clean inside as they are outside'.

Some farmers used to turn their sheep and lambs out in fields with wood-edges where this plant grew, moving them at least a month before they were due to be slaughtered for human food. But the Garlic during their period of growth was thought to be very beneficial to their condition.

Crow Garlic (*Allium vineale*), which is a common weed of poor pastureland, is also edible and no one will grumble if you wish to dig this plant out of fields so that you have an available supply – it should, however, only be used sparingly.

GLASSWORT   *Salicornia* spp.

Popular names: SALTMARSH or SALTINGS; LESSER PICKLE-PLANT.

If Glasswort were drawn and enlarged anyone could be forgiven for thinking that it was a succulent that had come from distant and desert regions. But in fact it grows in the mud of saltmarshes and often covers wide areas with its weird, sometimes branched, stems.

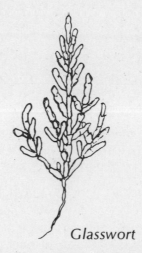

Glasswort

The curious, jointed, swollen branches, which are seldom more than six inches tall, make it look like a fun plant. It is a serious plant to botanists, however, and the only fun lies in seeing a group of them all clustered together, bent over to examine Glasswort through magnifying glasses to establish its exact identity.

This need not worry anyone who is after free food for they are all edible, as long as they are growing on unpolluted mud.

BEWARE of the soft, oozy places where you pick this plant, for they may look firmer than they are. Do not go Glasswort-gathering by yourself unless you know your area very well.

Always wash Glasswort well before either eating it raw while young and fresh, or lightly boiling it as it grows bigger and serving it with butter. Some people cook the plants whole, hold them by their roots and dip the hot green branches into melted butter flavoured with finely chopped mint.

As I am a reluctant pickler, I eat Glasswort green but it can also be 'put down' in jars, without its roots, in spiced vinegar.

If you are wondering at its English name as there is nothing really glassy about its tender green branches, it is called this because, in and possibly before the sixteenth century, it was burnt to ash and mixed with sand to use in the manufacture of glass. The then contempory herbalist Turner referring to it as *Glaswede*.

GOLDEN SAXIFRAGE *Chrysosplenium oppositifolium*

Country names: LADY'S CUSHIONS; GOLDEN CRESS; CREEPING JENNY (one of many others with this name).

Ditches and streams with steep, sheer sides are frequently lined with Leafy Liverworts (see page 78). In some places an almost equally low and green-leaved flowering plant grows among them. Amateurs at wild flower identification must study the leaves carefully unless this plant has its small yellow flowers showing, in which case it is easy to spot.

Golden Saxifrage leaves are eaten in mountainous districts of France as *Cresson de roche*.

GROUND ELDER   *Aegopodium podagraria*

Country names: BISHOPSWEED; GOUTWEED; DOG ELDER; DUTCH ELDER; GARDEN PLAGUE.

The last of the popular old country names for this plant describes its status admirably to most gardeners. Once in a garden, it seems almost impossible to get rid of it but the leaves can be cooked as a vegetable.

They should be picked off their stems and lightly boiled, like Spinach, and there is no doubt that the constant picking of leaves does, in time, weaken even this persistent horror. You may even reach the stage when, as I do, you have to go to someone else's garden for a culinary crop!

This plant has plenty of herbal history. It was regarded in the old days as a specific cure for gout and it may well have been introduced into gardens for this virtue alone. The name Bishopsweed follows on naturally, for according to some authorities Bishops used to live so well that they were frequently sufferers from 'the gout'.

Gerard knew it in the sixteenth century and said that it: 'groweth of itself in gardens without setting or sowing and is so fruitfull in his increase that where it hath once taken root it will hardly be gotten out again.'

There are plenty of places where Goutweed, Ground Elder or Bishopsweed grows on wayside banks, waste places and at the edges of woods, presumably where irate and tidy gardeners have tried to take it far from home!

GUELDER ROSE   *Viburnum opulus*

Country name: EPIPHANY TREE.
The berries of the wild Guelder Rose ripen either just as its leaves turn pink in autumn, or when the shrub or little tree is first bare of leaves altogether. Anyhow, they are very outstanding and show up in hedgerows like shining ruby droplets.

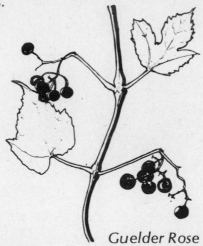

Guelder Rose

Mistle Thrushes, Song Thrushes, Blackbirds and incoming Redwings and Fieldfares eat them voraciously and leave very few for human pickers.

If you come across a good supply, try a few with stewed apple. They are rich in vitamin C and have been used as a substitute for Cranberries.

GUTWEED   *Enteromorpha intestinalis*

Have you ever seen groynes and pier foundations, at low-tide, covered with a bright green, shiny, wet-looking-grassy plant? This is Gutweed, a green seaweed, which grows profusely between intertidal reaches, on almost any surface from that of rocks, muddy shores, breakwaters, ship's bottoms, to the outside of any old cans which have been in the water long enough!

Looked at very closely, through a magnifying-glass, it can be seen to be finely tubular in construction, and with its irregular constrictions does resemble very small and thin intestines, or guts.

The peak growing period of this emerald seaweed starts soon after Christmas and attracts big flocks of migrating and hungry Brent Geese to our coasts, especially to graze on it in shallow, infrequently used harbours.

Gutweed

It is also edible to humans, but needs extremely careful gathering from places free of sewage outlets and other sources of pollution. Take a bagful home, if you find some that is untainted, wash it thoroughly, and cook it in a little boiling water, or in a steamer for only two to three minutes, according to taste. It is sometimes enough, when Gutweed is young, just to pour a kettle of boiling water over it. Start with very little at a time and when you are more used to the taste, increase the portion.

HAWTHORN   *Crataegus monogyna*

Country names: MAY; WHITETHORN; BREAD AND CHEESE TREE;
AZZY TREE.

The beauty of all the Hawthorns, showing up over fields,
hedges, woodland edges and beside the roads, always
make me wish that foreign visitors to Britain could arrive
in May when their flowers are at their best. Some of the
trees are so smothered in blossom that their leaves are
hardly visible. But May must not be picked and brought
indoors for floral decorations because it is traditionally
unlucky.

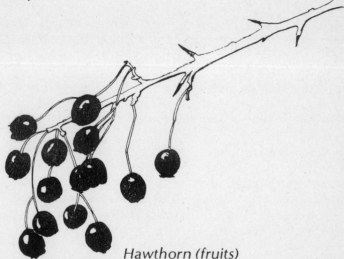

Hawthorn (fruits)

Country men and women, before pollution from
passing motor traffic, used to pluck and eat the young
leaves and flower buds as a spring tonic. They are not
actually as pleasing to the taste as they look, being slightly
tough even as they first come out, but they can be added,
chopped up, to spring salads.

The dark red berries which follow the flowers and are
ripe in autumn, have a variety of country names which
include the usual Haws, and Birds'-meat, Hogberries and
Pixie Pears. Some herbalists use an extract from them as a
heart-tonic.

As a culinary fruit I find them disappointing and
tasteless but a conserve can be made from their pulp

together with plenty of sharp-flavoured apples, as they are dry alone. Eaten raw, Haws are sweeter and of a better texture, I think, than cooked.

HAZEL   *Corylus avellana*

Country names: FILBERT-TREE; FILBEARD; COBNUT; WOODNUT; HALSE.

'Grow a hazel and a rowan, and a large patch of "Joy of the Ground" or periwinkle, near your door and you will never have witches in your house'! This is an old Sussex saying and was repeated in all seriousness to me when I arrived to live here in 1961.

Hazel, of course, is one of the providers of forked twigs for water-diviners and it has been regarded with respect because it was always purported to be the tree of knowledge, particularly in Ireland.

The nuts, which often grow in linked pairs, have long been recorded as symbols of fertility, like the curious somewhat testicle-resembling root-tubers of some wild orchids.

In Mediaeval Normandy the tree was allocated to St Phillibert because, it seems, this Benedictine Saint's feast-day, on 22 August, was when the Filbert nuts were thought to be ripe. It is a little early, however, and it is best to go gathering Hazel-nuts, Cobnuts or Woodnuts later in September.

HEATHER   *Calluna vulgaris*

Country names: LING; MISTFLOWER; DRY HEATH.

Although it would seem a sacrilege to pick the individual flowers off while they are at their honey-scented best, as well as being a finicky and endless performance, I am assured that people do do this and use the purple blossoms to make a fresh tea.

Sometimes Heather 'bells' are used to flavour teas made from other herbs. There is a story that Robbie Burns drank a 'Moorland Tea' in which Heather flowers together with those of Wild Strawberry and their leaves, and Bilberry, Blackberry, Speedwell and Thyme leaves were all used in the same mixture.

HEDGE GARLIC   *Alliaria petiolaria*

Country names: JACK-BY-THE-HEDGE; POOR MAN'S MUSTARD; PENNYCRESS-OF-THE-HEDGE.

A big collection of popular country and local names usually means, as it does in this case, that the plant has been known and appreciated for its uses for a long, long time. The above are only a sample of the picturesque titles of this common wayside weed.

The almost circular, penny-sized and rather dark green leaves of Hedge Garlic start showing early in the year and make the plant simple to recognize. As the leaves grow bigger and the white flowers also appear, the scent of garlic from it seems to me to increase.

Taste only a small amount of the young leaves, boiled in salted water as a green vegetable, to see if you like them (they can be distinctly bitter in flavour) before wasting others in preparing a big dishful. In any case, *never pillage hedgerows at all heavily for this or any other plant*, as they may be, like Hedge Garlic, a food plant of one of our native butterflies.

Female Orange-tips (the males only have upper top wingtips that look as if each has been neatly dipped into a pot of orange ink) lay their eggs on Jack-by-the-Hedge as well as on Lady's Smock, and would have difficulty in surviving if these wild plants were to be exterminated.

HEDGEHOG MUSHROOM   *Hydnum repandum*

Country names: URCHIN-OF-THE-WOODS; SPINY MUSHROOM.

This fungus earns its popular name because its spore-bearing layer, beneath the cap, is toothed, not gilled like true Mushrooms, or spongy like that of the Cep and other Boleti.

It grows in ferny woods and is pale velvety orange or yellowish on top with a frequently very irregularly-shaped cap.

Urchinsof-the-Woods can be good eating, but are better after a rapid preliminary boiling to remove their bitterness.

HERB BENNET   *Geum urbanum*

Country names: WOOD AVENS; GOLD STAR; BLESSED HERB.

Herb Bennet was the Mediaeval 'blessed herb', '*herba benedicta*' and 'herbe de Saint Benôit', which it is still called in France, and nothing at all to do with St Benedict as some writers have suggested.

It was often compressed and worn as a sacred amulet to keep evil spirits away and to ensure that 'venomous insects' did not come near.

This tall woodland plant has always had the virtue of being able to 'expel the Devil', or 'purge forth evill' and was once grown as a pot-herb in monastery gardens – chiefly, it seems for the practical reason of its clove-scented roots.

Gerard suggests that 'The rootes taken up in autumne and dried, do keep garments from being eaten with mothes, and make them to have an excellent good odour'. They are also, although they may not have a wide appeal, quite interesting to eat. At least I can say that they give an unusual flavour to casseroles and stews which, fortunately for the plant's sake, is not likely to become generally popular.

Culpeper thought that 'their roots had a delicate flavour', that should be 'used to comfort the Heart'. Modern herbalists use this plant as an astringent for digestive troubles, catarrh and as a general tonic.

Cosmetically speaking, Herb Bennet made into a *tisane* can 'clear the skin and remove spots and freckles' if taken internally and used as a face-wash externally.

HOGWEED   *Heracleum spondylium*

Country names: BROAD KEX; COWBELLY; COW PARSNIP; PIG'S-FOOD.

In view of the now well-known fact that the juices from the stems of the related **Giant Hogweed** (*Heracleum mantegazzianum*), which grows taller than a man, can inflict acute skin blisters on some people, I am very dubious about recommending this smaller plant as generally edible.

Giant Hogweed only appears to affect those with

sensitive skins, or those who are foolish enough to pick and use its hollow stems as pipes or pea-shooters. It is known as a phytophobic plant, the presence of its juice on the skin rendering it susceptible to rapid blistering from the sun, or even a bright light.

The common and smaller Hogweed is sought and apparently enjoyed by grazing animals and picked, as far as I know, perfectly harmlessly, but its Somersetshire name of 'Scabby Hands' could well be a reference to it too causing a rash, or even, indeed, a possibly now unrecognized virtue of it curing a spotty skin. It is impossible to rely on such local titles.

Be careful, anyway, how you use Hogweed, but I am assured that its young, thick, fleshy stems are very pleasant, and harmless, if boiled until tender and then strained and eaten with butter.

HONEY FUNGUS   *Armillaria mellea*
Popular name: BOOTLACES.

This invasive fungus is hated by foresters, and all tree-owners and gardeners because it is a rampageous tree-killer.

It is one of the most dangerous tree parasites for its black, bootlace-like rhizomorphs creep a long way underground from tree to tree and in this way keep on starting off fresh trouble.

But the above-ground fruiting-bodies, which are orthodoxly-shaped toadstools, are edible. Some people dislike their flavour, but they are worth a try and are easy to recognize with their scaly, golden or pinkish-tan tops. The gills, under the cap-tops, are veiled with a white membrane covering them while they are young, but this breaks as they grow and expand, and only persists as a shaggy ring round their stems. They grow in tight clusters, and can be fried like Mushrooms.

HOPS   *Humulus lupulus*
Country name: GREEN BINDWEED.

It is the flowers and their remains, or green 'cones', of this climbing plant that decorate a few hedges well away from the cultivated hop-fields. The flowers' remains hang from the fragile stems and show how high their parent plant can clamber up with the help of less weak-stemmed plants. They are usually eagerly sought for their beauty by flower-arrangers but if you find them first, home-brewed ale can be flavoured with them. They should be dried and added to your usual brew.

Hops

Young Hop stems, if there is an abundance of them, can be picked and cooked like Asparagus, but BEWARE of the somewhat similar-looking **Black** or even the **White Bryony** shoots which are both poisonous. The herbalist Gerard said, in the sixteenth century, that Hop shoots were 'more toothsome than nourishing', but they are worth trying when you are certain of their identity.

Hop tea, made by pouring a pint (550ml) of boiling water on to a handful, or less, of dried Hops, is a good sleep-inducer and many sufferers from Insomnia speak highly of a pillow stuffed with Hops as a cure for this distressing state.

What perhaps is not generally realized is that Hops belong to the same botanical family as Cannabis.

HOREHOUNDS
WHITE HOREHOUND　　　　　　*Marrubium vulgare*
BLACK HOREHOUND　　　　　　　　*Ballota nigra*

Country people used to call White Horehound 'Marvel' and sought it eagerly to make into a syrup with honey to help all their winter ailments. It grew abundantly in places among the chalk hills and downs and it has been suggested that the little town of Arundel, in Sussex, was once known as 'Horehound-dell'.

Unfortunately, it is now almost extinct as a wild plant in Britain, but it can sometimes be obtained from specialist herb-growers for the garden.

It should not be confused with the Black Horehound, sometimes called the 'Purple Dead-nettle', which is one of our commonest weeds. Its tall, leafy and violet-flowered stems crop up on waste ground in the middle of towns and cities, as well as by hedges and roadsides everywhere and it ought really to have a modern popular name of 'Dust-collector' designated to it! However, even Black Horsehound was once picked as an antiseptic herb to be used 'against ulcers and the bite of mad dogs'.

HORN OF PLENTY   *Craterellus cornucopoides*

At first sight this grey axd black horn-shaped fungus looks anything but edible. I am assured, though, by gastronomic friends that it is good and that it can be saved and dried for subsequent flavouring uses.

Horn of Plenty

Horns of Plenty are often common and easy to find in woods in the autumn. They grow in colonies among dead leaves, but must be picked when they are young and fresh.

HORSE MUSHROOM   *Agaricus arvensis*

These large and often 'ring-growing' mushrooms are no longer plentiful in downland pastures and, of course, many of the wide pastures of the chalk hills have now been ploughed up to grow barley or other profitable crops.

The caps of Horse Mushrooms are white, but they turn yellower as they age. The gills are never pink like those of **Field Mushrooms**, but a dingy grey.

If you are lucky enough to find some, they will make excellent eating but they are inclined to be more indigestible than Field Mushrooms and need longer cooking.

HORSERADISH   *Armoracia rusticana*

The old herbalists might have told you that Horseradish's prime function was that it provided an excellent vermifuge, or worm-expellent.

They could have added that it was a cough remedy as well as one of the eight bitter herbs used during the Passover.

It was Gerard who recommended the root of this plant 'stamped with a little vinegar put thereto', as a spicy asset to 'meates', to be 'commonly used for sauce to eate fish with and suchlike meates as we do mustard'.

Strangely enough, the long root, over which different authorities argue as to the best time for its digging, is rich in vitamin C. When eaten raw, after being finely grated, it has diuretic as well as internally antiseptic properties.

Horseradish is a common coarse-leaved weed of waste places, roadside verges and hedgerows. It has a spire of small white flowers early in the summer, and an occasional root – dug, I think, in the spring, which means you must look out for and mark the plant when you recognize it – can be harvested and stored, kept dry, to be used as a relish.

Some experts suggest that Horseradish's 'great heat' needs ameliorating with the addition of a little Basil, Oregano or Dill seed. Salads with cottage cheese flavoured with Horseradish are good for slimmers.

ICELAND MOSS   *Cetraria islandica*

As this book about edible wild plants contains notes on
Fungi and Seaweeds, I have included this one edible
Lichen, although it is only found in the extreme north of
this country. It is commoner in more northerly countries
of Europe and has always been sought and collected for
food.

This, of course, is the famous 'Reindeer Moss', grazed
by reindeer who are enabled to survive in arctic
conditions because of its presence.

I am told that the Laplanders perfected the technique
of extracting this lichen's bitter and purgative principles
by 'repeated maceration in water'.

Iceland Moss

When found, Iceland Moss can be made into
nourishing jellies or used for soups or broths or in milk. It
can also be dried and stored to be ground for flour.
Herbalists in Europe still use this famous lichen for
respiratory diseases and as a demulcent tonic; it is 'for
they who are emaciated after sustained illness', and
'particularly those whose weakness arose from the great
discharge of ulcers; for dysentery, diarrhoea and the
'hooping-cough'.

Iceland Moss is now imported and can be bought in
various health-food shops. To make a jelly for
convalescents, or for anyone who needs light
nourishment, wash a handful of the lichen in cold water
carefully. Strain, pour boiling water over it and strain
again. Then put Iceland Moss into pan with a quart of
water and simmer until dissolved. Add juice from two
lemons, flavour with mint and a scrap of orange peel

grated, and sweeten with honey. Let it lose some heat, then strain finally, and pour into small glasses or moulds until cool.

INKY CAPS   *Coprinus comatus*

Common names: SHAGGY CAPS; LAWYERS' WIGS; 'ENGLISH WILD MUSHROOM' (U.S.A.)

'Lawyers' Wigs' is really the most descriptive name for these common autumnal, or even later summer, fungi which come up suddenly after rain in colonies on green grass meadows or roadside verges.

They deliquesce, as they age, and drops of black liquid drop from the edges of their caps as they roll up and rot. Hence their most popular name of Inky Caps. This black liquid has actually been used as ink.

Inky Caps

Call them what you will, Inky, or Shaggy Caps, or Lawyers' Wigs, they make good eating provided they are picked when young and sliced and fried, when their flavour is delicate but delicious.

JEW'S EAR   *Auricularia auricula*

One of the most fascinating attributes of fungi is their propensity for growing in extraordinary and often grotesque shapes.

Jew's Ear fungi grow from elder trees, even when they are skeletal and dead, and are visible all the year round.

Jew's Ear

They are difficult to spot when the weather is dry, but directly the air is moist these gelatinous fungi absorb moisture and can be seen to be very ear-like. They are a pale pinkish-brown and their folds, convolutions and membranous substance may not look very appetizing, but they are perfectly usable, if lightly cooked, in soups, stews, casseroles, or even to enrich pâtés.

JUNIPER   *Juniperus communis*

Juniper, although really only describable as a shrub, is one of our three wild conifers, the others being Yew and Scots Pine.

Juniper

Juniper grows on dry hills, preferably on limestone, although also in Scottish pine forests, and has become rare during the past fifty years. It is exceedingly brittle, so that branches break from it and help its destruction, when cattle, sheep or ponies are grazing amongst the shrubs and possibly scooping out hollows by its roots for lying and sheltering in.

Gin used to be flavoured with Juniper berries, when the word *Juniperus* was changed to the French *genèvieve* and to the British 'Geneva' or gin. The berries are a powerful diuretic, and a few crushed and made into a herbal tea will soothe urinary discomfort. But they are also an abortiant and 'Kill-bastard' is one of this plant's most commonly found popular names.

A few berries, if the shrub is common in your area, are good flavouring for stuffings and forcemeat and they can be pickled in vinegar. They contain vitamin C and are an old anti-scorbutic remedy as well as being good, when steeped in boiling water and then sipped, 'to bring up the wind from the belly'.

KAFFIR FIG   *Caprobotus edulis*

This strange thick-leaved plant, brought to this country as a succulent from warmer climates for garden use, has become naturalized on cliffs and sandy banks in Devon, Cornwall and the Channel Islands.

It occasionally, especially during really hot summers, sets fruits, or 'figs' from its pink or yellow flowers which when ripe are quite pleasant and safe to eat.

KELPS   *Laminaria digitata*
and
L. saccharina

Popular names: TANGLES; FINGERED KELP; SUGARY KELP; SEA RIBBONS; SEA GIRDLES; MERMAIDS' GIRDLES.

Both these big, strong, brown seaweeds grow below the low-tide line and are frequently washed up on our shores. Storms tear them from rocks, sometimes with their roots or holdfasts still attached to pebbles, or fragments of rock.

The Fingered Kelp (*L. digitata*) has a thick gelatinous stem which branches into a wider, flatter and thinner 'hand', but the Sugary Kelp (*L. saccharina*) has only a short stem that quickly flattens into a very long, single ribbon-like frond with a frilled and fluted edge.

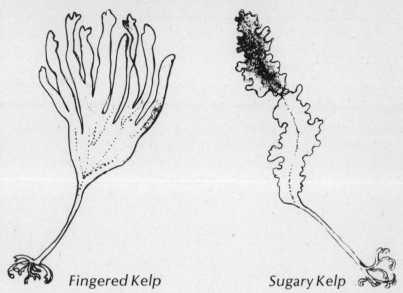

Fingered Kelp   *Sugary Kelp*

They have both been used for centuries as top dressings or manure for the soil as they are rich in potash and are very beneficial to the hungry land. They also have a long history of industrial use and in the eighteenth century they were collected to dry off in the sun before being burnt to ashes from which soda and potash were extracted to be used in the manufacture of glass and soap.

Iodine was discovered in 1811 and was obtained solely from these brown seaweeds.

Now, of course, they are important industrially for the alginates they produce, for which there is an enormous variety of uses.

Powdered, dried Kelp is a splendid addition to most peoples' diets. It supplies minerals that are not otherwise easily available. It is also good for dogs, cats and horses. Cattle and sheep near the coasts where Kelps are lying freshly brought in by the tide often choose to browse on them.

LADY'S SMOCK   *Cardamine pratensis*

Country names: MILKMAIDS; LADY'S MANTLE; LADY'S SMILE;
LADY'S GLOVE; PIGEON'S-EYE; PIG'S-EYE; CUCKOO-FLOWER.
It is also called the CUCKOO'S SHOES AND STOCKINGS; his
'EYES'; 'BEARD'; 'BREAD' and 'SPICE'.

It would seem better nowadays to leave this enchanting
wild flower to come into its lilac bloom among the golden
Marsh Marigolds and young spring grass than to pick its
early leaves for salad.

But if it grows in profusion near your home, it is worth
using as another member of the Crucifer or Cabbage
family. Do not pull it up by the roots, just pick off some
leaves to try as an addition to late winter diets, when they
are good in sandwiches and rich in vitamin C.

Lady's Smock

For my own delight (and I hope yours too) I have
included some of the different country names of this
popular and well-known English wild flower. Notice how
many 'Cuckoo' names it has accumulated over the years.
This is because the flowers are usually at their best when
this long-awaited bird should arrive in April. It is
interesting too that it is linked frequently with 'Our Lady'
and her sad association with Christ's death on Good
Friday and her happiness when he rose again on Easter
Day.

Beware of using popular names only for this flower. It
shares some of them with other less innocuous plants in

some localities. It is a food plant for Orange-tip butterflies (see Hedge Garlic, page 65).

LANDCRESS   *Barbarea vulgaris*

Country names: ST BARBARA'S CRESS; WINTERCRESS.

St Barbara, for whom this plant was named, has her Feast Day on 4 December when, providing the land is not locked in frost or covered with snow, it is perfectly possible that the dark, over-wintering leaf-rosette of this plant will be found in Britain.

Landcress (which is now cultivated and can be bought as seed for the garden) grows in waste places, or bare edges of fields, or even on rubbish-tips and if you find a colony, it is worth picking your daily quota *without* pulling up the roots. All green vegetables are better used as fresh as possible so that their vitamin content does not diminish.

The leaves can be chopped for salads or sandwiches, or if you have enough and they are very tough, lightly simmered for a few minutes in a little water, to make a dish of green vegetables.

Landcress, whether wild or cultivated, tastes very like Watercress and is gaining in popularity because it does not share the dangers of pollution, or of liver-fluke infestation like the water-growing **Watercress** (see page 118). But be careful where you pick it in the wild, and wash it well if gathered near roadside dust.

LAVER   *Porphyra umbilicalis*

Other names: PURPLE LAVER; SLOKE.

Most people have heard of 'Laverbread', a delicacy made from this particular red seaweed in Wales. Fortunately, Laver grows round other coasts as well and is not difficult to recognize.

Its fronds are dark red, almost purple, in colour and very thin and irregular in shape. I collect it into a polythene bag, take it home as fast as I can, wash it over and over again and then simmer it until it is a gelatinous mass.

Once cooked it is delicious coated in oatmeal or

wholewheat flour and fried in sunflower oil, or bacon fat.
Dried Laver is made into gelatinous cakes in China.

LEAFY LIVERWORT   *Marchantia polymorpha*

Popular name: COMMON LIVERWORT.

I had heard somewhere that this Liverwort, which
sometimes grows thickly along stream and ditch sides,
had in the past been extolled for its edible virtues but I
could find no references to either its old or possibly new
uses.

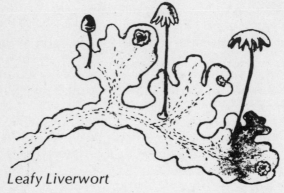

Leafy Liverwort

The British Museum (Natural History) replied to my
query that Liverworts, although they appear to be quite
harmless if eaten, have no proven worth to the epicurean
taste or to the pharmacologist. To quote from a
nineteenth-century hepatologist, Richard Spruce: 'It is
true that the *Hepaticae* (liverworts) have hardly as yet
yielded any substance to man capable of stupefying him
or of forcing his stomach to empty its contents, nor are
they good for food.'

Both the ancient writers on herbs, Dioscorides and
Pliny attributed medical properties to the Leafy Liverwort
Marchantia and probably saw in its lobed growth some
resemblance to the lobes of the liver. This idea could
have been carried on into the Middle Ages in the
Doctrine of Signatures.

The botanist from the British Museum went on to say
that as far as he was aware, no recent research had given
any support to these old suggestions.

It would be interesting to be able to gaze into a crystal

ball to find out if these widespread plants do indeed serve some purpose apart from providing sheer damp ground-cover.

LIME FLOWERS   *Tilia europaea*

It is the pale flowers of this beautiful tree that should be gathered on a hot day in full summer, just as they start opening. This is when they are strongly honey-scented and at their best to attract foraging bees to come for their nectar and thereby pollinate them.

Lime Flowers

Spread them out on sheets of paper to dry for future use, for making the delicious honey-scented *tisane 'tilleul'* which is used so much in France, even to soothe over-tired or fractious, teething babies. Finish off the drying process by putting the sheets of paper, once you are certain the flowers have no moisture on them, in a dark but airy place. Sunlight is too bleaching for them and if the air cannot circulate freely the Lime flowers will go musty. Leave until completely crisp and dry (which can take up to ten days), then pack lightly into screw-top jars.

Lime tea makes a quietening bedtime drink for anyone. It can always beneficially be sipped hot as a replacement for China or Indian tea specially by those who suffer from acidity or nervous indigestion.

LOVAGE *Ligusticum scoticum*

Common name: SEA PARSLEY.

Lovage is a handsome, bright and shiny-leaved Umbellifer which grows on some cliffs in Scotland, Scandinavia and Germany. In times of hardship it has been used as a vegetable with a flavour which most resembles Celery, but with a less pleasant lasting over-taste.

However it is an anti-scorbutic and contains vitamin C.

It is certainly not worth a dangerous cliff-climb to gather a leaf or two to taste, but if it is freely accessible it is worth tasting, or even trying the cultivated Lovage in your garden as a 'different' flavourer.

LUNGWORT *Pulmonaria officinalis*

Country names: SOLDIERS AND SAILORS; JOSEPH AND MARY; TINKER-TAILOR; CHANGECOAT-FLOWER; BLUE MARY.

Lungwort has great vitality and often after a plant has been thrown out in garden rubbish, it will flourish a long way from houses. There was a well-established colony in Surrey, just over a hedge where a passing motorist had emptied a sackful of weeds and unwanted plants, where I once used to collect a few young leaves each spring to chop for salads.

Lungwort

The leaves are hairy and spotted with irregular patches of white. They are a good example of the old theory which was known as the Doctrine of Signatures for they are more or less lung-shaped as well and so their appearance suggested a resemblance to the spotted lungs of a consumptive to the old herbalists.

'Like' being thought to 'cure like', they used to seethe (or simmer) a few leaves in milk and give the liquid to those who needed an emollient, cough-soothing drink. As Culpeper said: 'for all coughs, wheezings and shortness of breath', Lungwort was recommended to cure 'both man and beast'.

It is strange indeed, but the old magic still has some effect and modern herbalists often include tincture of Lungwort in cough cures.

MALLOW   *Malva sylvestris*

Country names: WAYSIDE MALLOW; COMMON MALLOW; FAIRY CHEESES; PANCAKE PLANT; RAGS AND TATTERS; BILLY BUTTONS.

If ever any wild flower earned the name of a ubiquitous weed, the Common Mallow ought to be one of the first to qualify. It grows along so many roadsides, on almost every vacant square yard of rubbish tips and on waste ground. It is what I call an opportunist, growing wherever it can, an equal, as it were, to the birds which follow man's buildings, the starlings and the house sparrows.

It is only seldom realized that this plant has any edible possibilities at all, although country children have always pulled the flat, round, cheese-shaped fruits and picked the seeds out of them to nibble.

The young five-lobed leaves are also edible but need very careful washing to remove all the dust and roadside pollution. Try them raw, and when they are older steam or boil them lightly when you will find them glutinous enough to add to soups or casseroles. They also go well with boiled brown rice, or to make any type of vegetable savoury, with plenty of flavouring and onions and tomatoes.

Common Mallow leaves have now been found to contain vitamins, A, B_1, B_2, and C.

Medicinally, this plant has been used as a digestive, diuretic and blood-purifying herb and also to make an external lotion for those with acne.

MARJORAM   *Origanum vulgare*

Country names: JOY-OF-THE-MOUNTAINS; ORGANY.

Wild Marjoram grows on chalk hills and in other lime-rich places. Its scented flowers are greatly sought out by bees and butterflies and its principal virtue, apart from this, is that like other members of the Mint family it contains the essential oil, Thymol.

This is, of course, a stomach tonic and a carminative-aid and a diaphoretic, or sweat-inducing herb. It can be used too for expectorant purposes, 'for getting rid of the phlegm'.

The usual culinary Marjoram (*Origanum Marjorana*) has a stronger flavour and is easy to grow. Its presence in the garden, or even in a windowbox, will save using the wild herb now that it and the situations in which it grows are less plentiful.

Culpeper, as so often happened, had a good use for it. He recommended that it should be 'made into a powder and mixed with honey, [when] it taketh away the black marks of blows and bruises if thereto applied'.

MARSH MALLOW   *Althaea officinalis*

Country name: VELVET-LEAF.

The delicious, chewy sweets known as 'Marshmallows' were originally made from a glutinous syrup obtained by boiling the roots of this attractive plant.

That was in the days when it was a common wild flower, growing particularly well on the East Anglian coast, as well as on similar shore-tops, and in estuaries on the Continent. But now it is a far from common sight to come across Marsh Mallow plants with their suede-green, soft and hairy leaves and transitory pale pink, Hollyhock-like flowers.

If you find a good colony, it is worth picking a handful of leaves to dry for making future poultices. Apply these, after very hot water has been poured over them and they have then been squeezed out by the usual towel-twisting method, to external boils, abscesses, or many-headed carbuncles, 'to draw the evil in them out to a point'. There is nothing better. Or you may prefer to use your handful of leaves to make a green Marsh Mallow drawing ointment. (See page 55 for directions for making Elder ointment and substitute Marsh Mallow leaves instead.)

Marsh Mallow

MEADOWSWEET   *Filipendula ulmaria*

Common Names: SWEET HAY; QUEEN OF THE MEADOWS;
HAYRIFF; KISS-ME-QUICK; TEA-FLOWER; COURTSHIP
AND MARRIAGE; DOLLOFF.

The creamy clusters of the sweet-scented Meadowsweet still decorate many railway embankments in high summer in rural England but they are disappearing fast from the old water-meadows where they used to make corners full of fragrance. This is mainly because so many low-lying fields are now being drained more effectively than they used to be, so that they can be used for crop-growing.

I still collect a bagful of flowers from such waste places as accessible railway banks, or by the river where there are plenty, to make a refreshing *tisane* which is excellent for curing mild indigestion or acidity.

The young leaves are useful for this purpose too, although they do not make such a fragrant drink. In fact, the fresh leaves and stems of Meadowsweet, when crushed, smell strongly of disinfectant and it seems likely that it was this contrast in scent between the flowers and the rest of the plant that led to its ironic country name of 'Courtship and Marriage'!

Meadowsweet

Once upon a time, Meadowsweet was one of the most sought after of all the herbs that were harvested and used with Bracken or the first-threshed straw for covering the floor, or strewing, before carpets and other floor-coverings such as we know now were in common use. The plant's strong antiseptic smell must have been welcome then, as was that of other strewing herbs like the Mints, Melilots, Lavender, Rosemary and Tansy, or even the smaller Sweet Woodruff, which was kept 'for my Lady's Chamber'.

MEDLAR   *Mespilus germanica*

Medlar trees sometimes grow in old country gardens and occasionally survive when the cottage and the rest of its garden has long disappeared. They are only very rarely found in a really wild state.

Their fruits are not ripe until late autumn, or even the beginning of winter, and there is no point in picking them until they look and feel 'past their best'!

When over-ripe the pulp can be scraped from the skin and core and eaten with 'thick cream and sugar'.

Medlars from Southern Europe are said to be far more

pleasant in taste as they ripen and soften earlier – before they begin, as our usually do, to go rotten.

MELILOT

Melilotus alba,
Melilotus altissima and
Melilotus officinalis

Popular name: WHITE OR YELLOW LUCERNE.

The Melilots, with their narrow spikes of small yellow or white pea-shaped flowers are becoming increasingly common weeds of roadsides and waste places. They are much loved and sought out by honey bees and bumblebees, and as you can see from their Latin names, one of them has been used as an 'official' medicinal and therefore edible plant.

Melilot

The leaves and seed-pods of these slender wild flowers are picked and used for making a substitute 'bean soup' in the United States and the young shoots are said to be pleasant as a vegetable. At least they are worth tasting.

A slight warning here, for this Melilot has been nicknamed 'Wild Laburnum' and the Laburnum tree is highly poisonous. So BEWARE of all parts of the Laburnum tree and do not get the idea that all plants with pea-shaped flowers are safe to eat.

All the Melilots smell delicious and cattle obviously find their taste pleasant. They contain coumarin, the substance which gives hay its typical 'scent', and which also has useful medicinal virtues.

MINTS   *Mentha* spp.

There are several useful wild and naturalized Mints that still grow in good supply, including the strong-flavoured Water Mint with its fat round, terminal heads of small mauve flowers.

Really it is a matter of taste as to which you use, but leave the small, creeping Pennyroyal growing if you come across it, because it has become rare now. Once upon a time it was in constant plentiful supply by each village pond because of its reputation as an abortiant (it was called 'bastard-killer', like Juniper) and, curiously enough, as a flea-killer as well.

Mint

The culinary uses of the Mints are too well known to go into here, except perhaps to mention that nowadays they are good for juicing, together with the leaves of other fresh green herbs, or can be put down for the winter in the freezer. For juicing, freezing, or even drying, the Mint leaves should be picked at their best and stripped from their stems. If you are drying them, spread the leaves out on paper on top of an eye-level grill, or somewhere similar. They should only get a warm air, no direct heat, but they stay greener if dried reasonably fast. (This rapid method is not recommended for other herbs, although I do my Parsley like it as well.)

Mints have endless medical possibilities and Mint tea is excellent served hot in the winter, sweetened with honey and given a pleasant tang by the addition of a slice of lemon or orange. Unsweetened Mint tea is good for settling nausea, even early morning sickness, and if used absolutely plain and dabbed on externally, will stop the maddening irritation of pruritus. (See Nettle, page 89, for this last purpose too.)

MOREL *Morchella esculenta*
Common name: FALSE TRUFFLES.

Truffles grow underground, but False Truffles, or Morels, have stems and grow three or four inches above the woodland floor, looking rather like a toad squatting on a thick, short, fluted column. They are very dark brown and their bluntly conical caps are convoluted and pitted so that they also look like ancient sponges.

Morel

Morels come up abundantly in the spring, in clearings where the ground has recently been disturbed or where there have been bonfires. Never, as with everything else, pick them all if you do find a group – leave some growing to produce spores for future crops.

These edible fungi can be used in a variety of ways. Their caps prove to be elastic and can be stuffed with savoury fillings and then baked. They can also be dried for future use by putting them in a *very* low-temperature oven for several hours, cooling completely and storing in dry and airtight containers.

MULBERRY   *Morus nigra*

The luscious dark red fruits of Mulberry trees are frequently wasted. Mulberries do not grow wild in Britain but are occasionally encountered in parks and old gardens where the collection of at least the fallen fruit would not be objected to.

'Mulberry Cheese' is a thick opaque jelly, delicious with meat and game as well as when used as a jam. It needs green apples to provide pectin as a setting agent and is every bit as good as redcurrant jelly and has a much more novel flavour.

I juice Mulberries with bananas to make refreshing, cool and very satisfying milk shakes. I also use them, again with bananas, to improve cottage cheese when I am tired of eating it in ordinary salad-dishes. If cooked and puréed with windfall apples, Mulberries freeze well.

Mulberry wine, made from ripe fruit, is a delightful winter drink. To make it boil 1gal (4.5l) of strained juice in 1gal hot water (or equivalent small proportions) with a little cinnamon for half an hour. Then add 6oz (175g) sugar and 1pt (550ml) sweet white wine. Cool and let the mixture stand for a week, strain, and store in a cool place in stoppered bottles.

Mulberry Rob (or 'cheese') is a recipe that is said to have originated in the Middle East, which conjures up pictures of dusky beauties with stained lips reclining on silk-covered divans, but which has been appreciated in cooler Britain for centuries too. To make it, boil 1pt (550ml) Mulberry juice with $\frac{1}{2}$pt (275ml) honey *very gently* until the mixture clarifies. This is not a good 'keeper', though so far in my household I have never had the chance to find out how long it would last. Store it carefully, however, if you hope to keep it.

MUSHROOMS   *Agaricus campestris*

Mushrooms are almost too familiar to need any description, and really too uncommon in most fields nowadays to confuse with any other white-topped toadstools. But BEWARE of eating others which look similar unless you are positive of your identification.

Field Mushrooms are surrounded by superstition and

folklore. You may have read that all fungi are safe for humans to eat if they have been nibbled by animals. This is rubbish, for slugs and some rodents appear to relish the pale green caps of the Death Cap and others that are nearly as lethal. It is also rubbish to say that all fungi that will peel are edible, or that those which, while cooking, do not turn a silver spoon black if it is inserted into the pan with them are safe to eat. *There is no safe way of identifying mushrooms except by sight and by knowing their botanical characteristics and appearance beyond all doubt.*

Once you are certain that you have found some Mushrooms you will doubtless know exactly what to do with them! Even if you have too many, those that are really firm and unexpanded can be put into the freezer for later use. An excess of 'buttons' are very good made into a mild chutney with vinegar and a few apples, or fried fresh in batter, like scampi, and frozen thus for keeping.

NETTLE   *Urtica dioica*

Country names: NAUGHTY-MAN'S-PLAYTHING; DEVIL'S LEAF.

Nettles are usually said to grow anywhere, but some soil experts say that their presence more often than not denotes the presence of a good soil. Folklorists used to say that their best colonies indicated that there was gold about!

Economically, their virtues are neglected nowadays. During the Second World War country women collected over ninety tons of them in Britain for their various County Herb Committes as a source of chlorophyll for medicinal use. But now clumps of Nettles are frequently sprayed in order to kill them in the easiest labour-saving manner.

If cut down before seeding, Nettles make first-class compost-fodder. Their young leaves can also be used as an excellent fresh green vegetable. I often put on a pair of rubber gloves in the early spring and go out and gather a bowlful. They need careful washing and a very quick boil in very little water to make a substitute for spinach. It is not worth sieving them unless you are making soup, for provided they are young enough they are very tender and their stinging hairs completely vanish while they are boiling.

Nettle-beer and Nettle-tea are old country spring-tonic drinks. It sometimes seems that less sophisticated people had more feeling for their bodies' needs than we do now, for they used wild herbs, apparently to good advantage, very freely. Of course there would not be the same numbers of cultivated vegetables and salad plants grown then.

Nettles were thought to be very beneficial as pick-me-ups after long winters and as the botanist, Sir Edward Salisbury, says, they would have provided vitamins and essential minerals to systems that were starved of fresh green food.

Culpeper advocated that Nettles should be eaten in Spring 'to consume the flegmatic superfluities in the body of man that the coldness and moisture of winter hath left behind'.

Chopped nettles mixed in raw with poultry- or even cattle-food are helpful, especially if the birds or animals are not running on grass.

Nettle juice makes a substitute for rennet for setting junkets, according to the late Mrs Hilda Leyel, a well-known herbalist who was deeply interested in plants and their uses.

A lotion made in the same way as Nettle-tea (approximately 1oz (25g) of herb to 1pt (550ml) of boiling water) is a soothing application for most forms of nettle-rash! It is good for burns, too, and sunburn and, in some cases, to quell the bad irritation of pruritus, but see Mint also for this purpose (page 86).

Nettles, by the way, form the food plants and the nursery diet of some of our most colourful butterflies, so leave some standing for the Peacocks, Small Tortoiseshells and Red Admirals to lay their eggs on.

NIPPLEWORT   *Lapsana communis*

Country names: CARPENTER'S APRON; BREASTWORT.

Nipplewort grows in all kinds of neglected corners and is a true 'waste-ground weed'. Its appearance is slender, though tough, and its wiry stalks do not contain milky sap like those of the in many ways similar-looking **Wall Lettuce** (*Mycelis muralis*).

The leaves of both these plants can be picked to use as salad additions or in sandwiches.

OATS   *Avena* spp.

Apart from the possibilities of gleaning on the edges of cultivated fields after the crops have been cut and carried, which is still permissible in some areas and can result in small personal harvests, it is quite unpractical to think of gathering the grains of the unwanted Wild Oat to grind for flour.

Oat

Wild Oats are a menace to farmers. Their seeds have enormous survival potential and germinate in all kinds of crops. They are virtually ineradicable unless they are pulled out by hand, and seem to spread rapidly. In huge fields in such areas as the Cotswolds, this hand-pulling is a laborious task and year after year some Wild Oats still go on surviving.

A few plants of Wild Oats are worth collecting and drying before they shed their grain. This grain, which is small in comparison with that of cultivated Oats, makes a splendid and soothing tea which is helpful to the sleepless as well as being a tonic for convalescents.

ORACHE   *Atriplex patula*

Orache is a common garden weed which even those who are usually thought to know a lot about wild flowers often avoid naming! This may partly be due to the fact that several of these all-green weeds are confusing but is also, I am sure, because no one ever knows how to pronounce it. I call it 'au-raitch', with a rolling 'R', and no one has corrected me yet!

The Common Oraches are variable in leaf-shape and one of them, the Spear-leaved Orache (*Atriplex hastata*), is very like **Dungweed**, so it is not surprising that they share *Chenopodium album's* old popular names of 'Muckweed', 'Dungweed' or 'Goosefoot'. All are safely edible and were once in frequent use as pot herbs, but BEWARE of their roots. Use only the green leaves as the roots contain saponins which can have a toxic effect.

OYSTER MUSHROOM   *Pleurotus ostreatus*

Popular name: VEGETABLE OYSTER.

Strictly speaking, this fungus is hardly mushroom-shaped. It grows out of tree-trunks, usually preferring Beech, and has irregular and somewhat bluish tops to its caps, with inrolled edges. The gills are white.

Oyster Mushrooms

Oyster Mushrooms must be picked while young, before their cap-colour changes to brown, fawn or a yellowish cream. They make delicious eating and their season, during the autumn, is longer than that of most fungi as they seem able to withstand light frosts.

PARASOL MUSHROOM   *Lepiota procera*

It is the height of Parasol Mushrooms that makes them so eye-catching. They tower over short downland grass, or pasture cropped uniformly low by grazing sheep, cattle or deer, and as they grow in grassy places in late summer they are usually particularly visible.

Their slightly conical and umbrella-shaped caps are distinctive too. Once fully expanded they are adorned by the dark brown scaly remains of the veil that covered the whole cap while it was immature, and it is possible to find one with a diameter of at least eight to nine inches.

The somewhat slender stem is circled under the cap by a movable ring, which adds, of course, to the plausibility of its English name.

Parasol Mushrooms

Parasol Mushrooms are edible and good if they are fresh, young and tender, and can be used for flavouring in all the same ways as Mushrooms. They can also be dried for winter use. But BEWARE, go slowly and not greedily with your first picking for there are records of upset digestions among those who are allergic to them. *And* do not try any other species of *Lepiota,* even if they look like *L. procera,* for there are a couple of smaller species which are extremely dubious.

PARSLEY PIERT   *Aphanes arvensis*

Common names: BREAKSTONE; FIELD LADY'S MANTLE.

It is only worth looking for this low, all-green herb on dry and more or less bare soils. In some places where conditions are just right for it, it grows in great quantities, but seldom reaches more than three inches in height.

Culpeper suggested that it made a good salad plant and that it should be pickled by 'the Gentry', in the same way 'as they pickle up Samphire for their use all the winter'.

As Parsley Piert chooses to grow on fast-draining and often stony ground, an infusion from the whole small plant was used for clearing the 'stone' in sufferers with urinary troubles. 'Our herbe women in Chepeside knew it by the name of Parsley Breakstone', according to Gerard.

Medical herablists still recommend an infusion of this plant for anyone suffering from urinary 'hold ups' or discomfort.

PENNYCRESS   *Thlaspi arvense*

Pennycress is another member of the Cabbage, or Crucifer, family which grows freely as an annual weed and makes a change in a salad, or can, like the **Bittercresses, Landcress** and **Watercress,** be put into sandwiches.

It grows in gardens, along field edges and on waste ground, and can quickly be recognized by its small white flowers which are rapidly succeeded by flat, green, winged fruits that get bigger and bigger until the oldest, at the bottom of the spike, are are least as wide as a twopence piece.

PINE   *Pinus strobus*

Common names: WEYMOUTH PINE; WHITE PINE.

The seeds of our common wild **Scots Pine** (*Pinus sylvestris*) are very small and hardly worth collecting to eat, but those from the 'imported' Weymouth Pine are bigger and tastier. This tree is frequently found in parks, and was named popularly after Lord Weymouth who planted many at Longleat in the early eighteenth century.

One way of identifying this particular pine is to look at a leafy branch, when you will see that the needles are in close bundles of five. Another is to search for the long, very resinous and narrow, slightly curved cones which have been blown down from the usually high crown of the pine. The cones only expand to release their paperwinged seeds when they are dry.

The New England Indians used to make the fresh growth of the leafy branch tips of this conifer into spring tonic drinks. The leaves have since been analysed in America and have been found to contain a high percentage of vitamins A and C. In fact, Ben Charles Harris in his American book, *Eat the Weeds,* says that 'Pine needles contain about five times as much Vitamin C as is contained in lemons', and that the Russians use an infusion of Pine needles as a source of this vitamin. Apparently the American Indians also had an edible use for the pollen.

The seeds from the **Monkey Puzzle Tree** (*Araucaria araucana*) are also edible and nourishing.

PLANTAIN   *Plantago major*

Common names: RATSTAIL; CUCKOO'S BREAD; SNAKEWEED; HEALING-BLADE; CANARY-SEED; WHITE MAN'S FOOT.

Plantains grow all over the world. In Australia, New Zealand and parts of North America this plant is called 'Englishman's Foot', because it seems as if it has appeared wherever Englishmen have trodden!

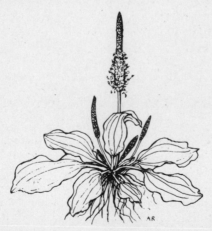

Plantain

The tall thin spikes of inconspicuous flowers and fruit are sometimes curved or, if any man's foot has actually damaged them by treading on them as they grew, they can be badly contorted. Hence, partly, the popular name of 'Snakeweed', although Ratstail Plantains have a very old reputation for curing 'the bites and stings of serpents and scorpions' as well.

Poultry when turned out into open runs, especially if they are green-hungry, go for Plantain leaves and crop them down. The leaves can be picked for human consumption and added to stews, soups and any green juicing mixtures, or even cooked alone like Spinach, but to us this common plant's chief virtues lie more in its medicinal properties.

Tea made in the usual way from approximately one teaspoonful of fresh or dried Plantain leaves over which a pint (550ml) of boiling water has been poured, has blood-clearing and diuretic properties. It is also credited, more in some parts of the country than in others, with the reputation of being able to cure anyone of the desire to smoke.

Plantain tea has been used as a cure for toothache, when it should be held *hot* in the mouth.

Fresh leaves crushed to release their scant juice make a quick first-aid soother for insect bites and stings. The juice also acts as a styptic for superficial but bleeding thorn-pricks, grazes and scratches. This is interesting because modern chemical analysis shows that as well as being rich in vitamins C and K, Plantain leaves contain factor 'T', which is known to be helpful in arresting bleeding.

PRIMROSE   *Primula vulgaris*

Country names: EASTER ROSE; SPRING'S SWEETHEART.

While conservationists everywhere are trying to make the public understand that by yielding to their personal greed and picking great bunches of such wild flowers as Primroses or Cowslips, they are only speeding up the time of their extinction (a picked flower cannot set seed), it may seem foolish even to mention that a good ointment for broken chilblains used to be made from Primroses.

But, of course, if you come across a multitude of Primroses and can bring yourself to do so, it is perfectly plausible to pick a few. Try eating a fresh Primrose flower, or putting one or two on thin brown bread and butter to tempt a child who will not eat salad or cooked vegetables.

Primrose

Incidentally, and I write as an ardent conservationist, Primroses are everyone's idea of the epitome of spring and the coming of hope and sunny days once more after long dreary winters, so that if you have a housebound, country-loving relative, neighbour or friend, I cannot believe it matters taking a *few* for them, to refresh their courage.

PUFFBALL   *Lycoperdon giganteum*

There is one giant edible Puffball which can grow bigger than a football, and is delicious if picked at the stage when it is young and covered with a white 'kid-leather' skin, then sliced and fried.

It comes up in late summer, after rain, in woods, fields, and even in rough corners of gardens, often among a bed of Nettles, and is unmistakeable.

This extraordinary looking Puffball was gathered and dried in the old days for tinder, and in the sixteenth century Gerard said that people also used it 'to kill or smother their Bees when they would drive the Hives and bereave the poore Bees of their meate, house and combs'. In those days, of course, there were no permanent artificial bee-'houses' such as we have now and the bees had to be killed, or rendered insensible when their honey was taken each year.

There are several species of smaller Puffballs which are also edible while they are young.

Puffballs

BEWARE of a globular, warty, somewhat yellowish and hard but in a way similar-looking fungus, the **Earthball** (*Scleroderma aurantium*), which can upset some people badly if eaten by mistake.

RASPBERRY   *Rubus idaeus*

Colonies of naturalized or 'wild' raspberries are frequently found in the country. They may originally have been started up by birds carrying the seeds from garden fruit, and then evacuating them after they have digested the sweet red pulp, or even by humans when they have picknicked and spilt a few during their feasts. But Raspberries can be truly wild plants as well.

The fruits are smaller and sweeter than those of cultivated Raspberries and take time to gather, but it is well worth the effort. Do not go crashing through their canes regardless of everything but filling your basket or you will damage all the plants that are in your way and incidentally spoil next year's growth, leaving the place looking as if a herd of elephants has just plundered it. (This also applies to Blackberrying!)

Wild Raspberries can be eaten raw or made into jams and can be kept successfully in the freezer.

Young Raspberry leaves gathered and dried make one of the best herbal teas for pregnant women. The tea acts as a tonic to the uterine muscles and should be drunk three times a day during the last few weeks when it really works in expediting the birth of the baby.

RED LEG   *Polygonum persicaria*

Country names: REDSHANK; LAMB'S TONGUES; PIGWEED; LOVER'S PRIDE; DEAD ARSESMART.

BEWARE of **Water Pepper**, or Arsesmart, (*P. hydropiper*), similar in appearance but usually smaller more slender, growing more by ditches and dykes and where standing water has dried out. It is not dangerous but its leaves are bitingly 'hot', hence the common and apt name of Arsesmart is true! It makes your mouth and throat smart too, if you eat it.

Red Leg

If doubtful, nip the end of a leaf of Red Leg gently and wait to see if it burns your tongue before picking any more. In any case, although it has the reputation for being edible to humans, go carefully with this plant as well.

Red Leg or Pigweed is a rampant farm weed, and is often said to be a sign of poor ground: it is recognizable because of its persistence, its 'knotted' joints which look swollen, the tufts of small pink or flesh-coloured flowers, and its narrow, lanceolate black-blotched leaves.

The blotches may be absent from the leaves, in which case the Devil has not got round to pinching them yet!

The plant is related to **Bistort** (see page 23) and the **Sorrels** (see page 110), and a few of its leaves can be eaten raw, or included in green juicings.

RED-SPURRED VALERIAN   *Centranthrus ruber*

Country names: BOUNCING BESS or BETSY; DRUNKEN SAILOR; CATBED; PRIDE OF PADSTOW.

A new popular name of 'railway-cutting Plant' could now be added to the many that this rapidly spreading colonizer of so many unpromising walls and cliffs has won since it arrived in Britain in the sixteenth century.

Its pink, red, or occasionally white flowers are very attractive to nectar-seeking butterflies and moths, as they are slightly fragrant, and it is one of the best plants to watch for the rare immigrant Humming-bird Hawk moths or to see hundreds of incoming Silver-Y moths when they congregate and hover over its flowers on warm summer evenings.

The first tender young leaves of Red Valerian are edible (when the plant is plentiful) in salads and sandwiches, and can also be lightly boiled as a green vegetable, but taste one before picking more, as some people find them too bitter.

ROSES   *Rosa* spp.

Surely the various scarlet fruits of the different wild briars or roses, the 'hips' in the old partnership of 'hips' and 'haws' (see page 63), are too well known to need a description. That they have been used since the Second World War as a valuable source of vitamin C must also be known by most people.

However, perhaps the use of the wild rose, or indeed some garden-rose petals is not so well recognized. They add colour and piquancy to different dishes, especially if you find the deep pink flowers of the Sweetbriar which grows in chalky and limestone areas. Rose petals can be decorative in salads and can also be made into petal jams, jellies, honeys and syllabubs, or used to flavour Turkish Delight. Rose-petal tea is even given to arthritics by herbalists.

The first rosaries were made from rose petals, when they were handled and collected and fingered and then pressed into bead-like shapes.

Personally, I leave the making of Rosehip jams and jellies and syrups to more professional cooks, as their lack of pulp and the hairy covering of their 'stones' always prove such a hazard to me! But there is no doubt at all of their high vitamin C content, which is said to be sixty times more than that of lemons, plus vitamins A and P as well.

Rosehips

Ripe wild Rosehips can be picked and their outer red skins and thin layer of pulp chewed by adults and children. If you are determined to have a go at the syrup, here are the 1943 Ministry of Food's instructions:

Have ready 3 pints of boiling water, mince the hips in a coarse mincer, drop immediately into the boiling water or if possible mince the hips directly into the boiling water and again bring to the boil. Stop heating and place aside for 15 minutes. Pour into a flannel or linen crash jellybag and allow to drip until the bulk of the liquid has come through. Return the residue to the saucepan, add $1\frac{1}{2}$ pints of boiling water, stir and allow to stand for 10 minutes. Pour back into the jelly-bag and allow to drip. To make sure all the sharp hairs are removed put back the first half cupful of liquid and allow to drip through again. Put the mixed juice into a clean saucepan and boil down until the juice measures about $1\frac{1}{2}$ pints, then add $1\frac{1}{4}$ lbs sugar and boil for a further 5 minutes. Pour into hot sterile bottles and seal at once. If corks are used these should have been boiled for $\frac{1}{4}$ hour just previously and after insertion coated with paraffin wax. It is advisable to use small bottles as the syrup will not keep for more than a week or two once the bottle is opened. Store in a dark cupboard.

Hedgerow Harvest, MoF, 1943

ROWAN   *Sorbus aucuparia*

Popular names: CHICKEN-BERRIES; COCK-DRINKS; CARES; QUICKEN-BERRIES; MOUNTAIN ASH BERRIES.

The orange late-summer-ripening fruit of the Rowan have earned themselves a strange variety of popular local names over the years, a few of which are given above. They are even 'Poison-berries' in some parts, which must mean that someone once ate too many raw, for they are all right for most people when cooked.

They make excellent jelly to eat with game, poultry, mutton or lamb, but need the addition of green apples, or crabs (which were used to make old-fashioned 'verjuice'), to make the jelly set.

SAFFRON MILK-CAP   *Lactarius deliciosus*

Country name: VEGETABLE SHEEP'S-KIDNEYS (France).

If you see a ring or a scatter of deep orange, somewhat sticky-capped toadstools growing under coniferous trees in a pine wood, look at them carefully before turning away in horror. Saffron Milk-caps are quite a delicacy and eagerly sought by fungus-eaters!

Saffron Milk-cap

Their caps are ringed in zones of shadings of green and

they also have curled-under edges. When the cap is broken, saffron-coloured 'milk' exudes from the break.

Young specimens of Saffron Milk-caps should be eaten as soon as possible after picking – wash them very well, then fry or simmer in milk. Some can be dried for cooking when they are out of season, but they do actually grow for three or four months from September onwards.

ST GEORGE'S MUSHROOM   *Tricholoma gambosum*

St George's Day is on 23 April, so this is a spring and summer fungus which grows, often in rings, from old chalky pastures when the weather is damp enough. It is a tall, handsome, mushroom-like fungus with big creamy or slightly ochreous caps and stems and white gills. The whole plant smells and tastes of strong meal when raw.

St George's Mushrooms are nothing like as common now as they were even twenty years ago, as so many of the fields and downland slopes where they used to grow have been cultivated.

Pick any that you are fortunate enough to find but only if they are young. Cook them like other mushrooms, but a little more carefully and for longer, or they may be tough.

SALAD BURNET   *Sanguisorba minor*

I have often found the many-leafletted leaves of this ground-hugging plant thirst-quenching when I have been out, without a drink, on a long hill walk. They taste faintly of cucumber and can be used in sandwiches or chopped up in salads.

They can also be used for a refreshing herbal tea which makes a good tonic and mild diuretic. Country girls made 'lotion' to add to face-washing water by soaking the leaves of this plant.

The flowers are a help in recognition and grow on tall (up to ten-inch-high) stems in globe-shaped masses as big as marbles. Though they look brown, closer investigation shows that each small flower has no petals, but green sepals, and that the upper flowers have red styles, whereas those lower down have produced yellow anthers.

SAMPHIRE — *Crithmum maritimum*

Popular names: SAMPER; SEMPER; SOFT-ROCK; *Saint Pierre* (France).

Samphire is not to be confused with **Glasswort** (see page 59) or **Kaffir Fig** (see page 74), although both are succulent plants. Samphire is a thick and fleshy-stemmed, yellow-flowered Umbellifer which grows on cliffs, or from sand near the sea, or on rocks, and looks, before it flowers, almost comparable with a succulent cactus with longish stems.

It is not worth risking your neck to get at Samphire if it is growing beyond your reach (although the 'samphire-pickers' in history often seem to have done this), for more than likely you will come across some that is easily accessible at the foot of the cliffs. Certainly in West Sussex, which is cliffless, it grows from sand and shingle.

It is best picked in early summer when fully-grown but not tough. Taste it before picking much, and in any case do not pick much from one plant. If you like the taste, pick it sparingly to go with salads and for trying as a sauce or as a green vegetable, especially with fish or vegetarian dishes.

Traditionally, it is usually pickled. Here is an old recipe for Pickled Samphire:

> Put Samphire green into a pan and sprinkle in the salt. Cover it with water and let it soak for a day and a night. Put the pan on heat, add more salt and enough vinegar to cover it all over. Simmer till green and crisp. Bottle it up in tightly lidden jars.

SCURVY-GRASS — *Cochlearia officinalis* and *Cochlearia danica*

There are several species of Scurvy-grass but most of them grow in coastal and estuarine areas, but they can also be found in wet mountainous conditions. Perhaps the 'official' herb, *C. officinalis*, is the commonest.

('Official' herbs were those that were, or indeed sometimes still are, used medicinally, or 'officially', and they frequently include the description in their scientific names, as does Scurvy-grass, *Cochlearia officinalis*.)

Everyone now knows the sad story of scurvy-suffering

sailors who were deprived of vitamin C on long voyages far away from land, and their discovery of the now famous Scurvy-grass which helped them so much. It was recorded by Gerard in the sixteenth century, and he boosted it tremendously by his descriptions after it had cured sailors of 'this filthie, lothsome, heavie and dull disease', 'in which the gums are loosed, swolne, and exculcerate: the mouth grievously stinking' and 'the thighes and leggs are withall verie often full of blewe spots, not much unlike those that come of bruses; the face and the rest of the bodie is often of a pale colour; and the feet are swolne, as in the dropsie'.

It was no wonder the herb-women did a brisk trade in their Scurvy-grass sales after that! And its popularity went on until the pleasanter-tasting juice of oranges and limes was understood to fulfil the same purpose.

The leaves of Scurvy-grass, particularly those of the very early-appearing Danish Scurvy-grass, C. *danica,* which is frequently found on shingle beaches and which has lilac-coloured flowers, can be used in salads to provide extra vitamin C, especially if you cannot eat oranges.

SEA BUCKTHORN   *Hippophaë rhamnoides*

The orange berries of this shrub are very rich in vitamin C and European herbalists use them to make a spring revitalizing tonic. Make as for Rosehip syrup (see page 101), but omit the sugar, and when using the syrup, add a little honey to sweeten it.

There seems to be no reason why a conserve should not be made from these berries if enough are available.

SEA CLUB-RUSH   *Scirpus maritimus*
Popular name: SALTMARSH BULRUSH.

It is saddening to read that in the eighteenth century the English poor were forced to use this plant a great deal in times of scarcity for flour for their 'breads, broths and soups'; for, although very handsome in its close green ranks at the edges of marine coastal waters, it does not look particularly edible or appetizing.

It was the roots that were used, and they were dug, dried and then ground to a power. There is another old claim, from America, that the Indians from the north-west ate the tender part at the root-end of each stem fresh.

Be careful if you try this plant to take it from unpolluted water.

SEA LETTUCE   *Ulva lactuca*

Common names: GREEN LAVER; GREEN SILK-SEAWEED.

When I was a child and green oiled surgical 'silk' was used to cover wet bandages, I always used to think that *Ulva* looked exactly like it. This seaweed grows in rock pools or from any rock, breakwater, pier or harbour surfaces that are sea-covered at high tides. Its fronds are flat, very thin and irregularly-shaped, with wavy edges and it can be found in sheets of up to two or three inches at its widest part.

Sea Lettuce

This green seaweed is a good dietary addition, providing chlorophyll and traces of minerals, such as iodine, that are needed by our bodies. But again try only a little at first until you are used to the unusual taste and texture.

Pick from the most unpolluted shores and cook, after careful washing, in a steamer. It is quickly reduced to a gelatinous pulpy mass which is delicious with boiled brown rice or macaroni cheese.

SEA PURSLANE *Halimione portulacoides*

Popular name: GREY-MAT.

Acres of this grey- or silver-leaved plant fill the edges of old silted-up unused harbours, inlets and saltmarshes and remain unremarkable to look at except perhaps to artists, or until each plant produces small branched spires of dark gold but insignificant flowers, or even later in the year when the whole mass turns to a dull pink colour in the autumn.

Sea Purslane

The leaves are edible and pleasant raw in salads, but need careful washing and, of course, must be picked from unpolluted areas.

SERVICE TREE *Sorbus torminalis*

Country names: CHEQUER, or CHECKER; CHEQUER- or CHECKER-TREE.

The berries of both this and the related **Whitebeam** (*Sorbus aria*) are beautiful to look at and edible but are very acid and not fit to eat at all, even cooked, until they are over-ripe like Medlar fruits. They contain vitamin C, which can, however, be more palatably obtained from other wild fruits and they should be left unpicked by humans, to satisfy wild birds' requirements.

In Gerard's time, the fruits of the Chequer-tree were used medicinally for 'settling the colick'. But **Bilberries** (see page 23) are also excellent for this and usually far easier to obtain. It was a modern physician, Dr Dorothy Shepherd of Harley Street in London, who seems to have started prescribing Bilberries for acute diarrhoea and dysentery in medical practice again, although herbalists have extolled their use for this purpose for centuries.

SHEPHERD'S PURSE   *Capsella bursa-pastoris*

Country names: LADY'S PURSE; HEN AND CHICKENS; MONEY-BAGS; BAD MAN'S OATMEAL; NAUGHTY MAN'S PLAYTHING; ST JAMES' WORT; PICK-POCKET; PEPPER AND SALT.

The 'bad' or the 'naughty' man in popular country names for plants and creatures was of course the Devil. His name was often used as an epithet for very common weeds – and nothing, surely, could be a more common field and garden weed than Shepherd's Purse.

The purse-like fruits were also often part of the plant's descriptive names – for instance, 'Money-bags' – but among the long list (of which the above are only a few) none seem to suggest that all the green parts of this rapidly-growing weed are edible, even if rather peppery in taste.

Medicinally, Shepherd's Purse leaves are anti-scorbutic and can be used to make a mildly diuretic tea. Used raw and chopped, small quantities of the leaves improve coleslaw.

SILVERWEED   *Potentilla anserina*

Country names: BREAD AND BUTTER; TRAVELLER'S-EASE; PRINCE OF WALES' FEATHER; GOLDEN SOVEREIGNS; GOOSE TANSY.

Silverweed, if looked at carefully, is one of our most beautiful common weeds. The clear yellow single flowers are set off to perfection by the feather-like silvery leaves, the fine crimson stems of the flowers and all the runners that shoot off from the ground-hugging parent plant.

It is not surprising that this plant has earned itself so many descriptive country names, a few of which are mentioned above, for Silverweed, as history shows, has been useful for food at times of famine. Then people disregarded the plant's beauty and ate its roots 'roasted or boiled or ground down into a kind of floure'. They were 'eaten eagerly' in parts of Scotland, where Silverweed was recognized as supporting 'the inhabitants for months together during a scarcity of other provisions', especially 'in seasons where their crops succeed the worst, when the roots of the "Moors" [as the plant was often called in the North] never failed to afford a reasonable relief'.

Silverweed

Try Silverweed roots now if you have an available supply and can bring yourself to dig them out. Use them as you would Parsnips, or for flavouring soups, stews or casseroles.

SLOES   *Prunus spinosa*

Country Names: HEGGS, or EGG-PEGS; BLACKTHORN-BERRIES; WINTER-PICKS.

Sloe fruits are like miniature black, but blue-bloomed, plums. They often persist, although they are inclined to shrivel, all through the winter, hence their name 'Winter-picks'.

Blackthorn or Sloe bushes have apparently always been surrounded by folk-lore, legend and superstition. Branches of the shrub's small, white, densely clustered flowers are, like Hawthorn or May, unlucky to bring indoors.

The flowers have been credited with diuretic and mild

laxative powers if used to make a herbal tea and the ripe fruits provide a good supply of vitamin C and make a stomach tonic for those who are 'too loose'. The old herbalists said that 'the juice of sloes do stop the belly, the lashe and bloody fluxe'.

Modern European herbalists use the berries to provide a vitamin C-containing syrup.

'Blackthorn Winter' in Sussex, when the hedges are as white as if it had been snowing, usually starts towards the end of March and is invariably a bitterly cold time.

SORRELS   *Rumex acetosa* and
Rumex acetosella

The shiny, flat, arrow-shaped leaves of Common Sorrel are easily seen in the spring among the narrow blades of meadow grasses. As a salad plant it is highly acid and has a good vitamin C content, but sufferers from all forms of acidity, including stomach ulcers, rheumatism, arthritis, and some forms of cystitis, should avoid it like the plague.

Sheeps-Sorrel (*Rumex acetosella*) is a smaller plant which grows on heaths and other sandy and acid and bare ground, the leaves of which can also be used, but the same warnings apply to it!

Both these plants provide leaves for Sorrel soup. Cook like spinach with very little water for a minute or two, or until tender, strain and put to one side. Cook an onion in milk, add pepper and salt, thicken with flour, and boil until creamy in consistency, flavour with a sprinkling of Marjoram or dried Oregano, stir in strained Sorrel and served with *croûtons*.

Sorrel is useful for juicing with other green herbs, or for chopping finely into sauces (which used to be kept entirely for 'rich fish').

The tea made from Sorrel leaves was once used only as a fever-reducer, or febrifuge, but is a refreshing, appetite-creating drink.

SPRING BEAUTY   *Claytonia perfoliata*

Claytonia is a small, white-flowered weed with clear-green twin top leaves that encircle the stem at the bottom of the flower-spike. It has only been in Britain since the middle of the last century, but it can spread very fast and cover flower-beds, or sandy banks or dry corners of fields.

Spring Beauty

It seems to avoid alkaline or chalky soils.

The whole plant, picked before it flowers, washed thoroughly and lightly boiled, makes a 'different' green vegetable.

STAR OF BETHLEHEM   *Ornithogalum umbellatum*

It may seem a sin to suggest that the bulbs of this beautiful flower, which is not thought to be a native of Britain but merely a garden escape, are edible.

Star of Bethlehem has been naturalized in this country for a long time and it spreads fast. It is easy to imagine cottagers chucking out clumps when their gardens got too overcrowded with it. Now many woods and hedgerows and copse-edges seem to have added it to their natural flora.

The white-striped Crocus-like leaves appear very early in the year and the 'Nap-at-noon' flowers, sometimes also called 'Doves'-dung', are out in May. Actually they are more often shut than open for as well as shutting at midday, they do not open at all unless the mornings are sunny. When they are open they look like bunches of little white stars.

In the U.S.A. it has been stated that this plant is

poisonous to cattle, but the bulbs are still dug up and eaten by humans there!

SWEET CICELY   *Myrrhis odorata*

Country names: WILD ANISE; SWEET BRACKEN; SWEET FERN.

Sweet Cicely is an easy-to-recognize Umbellifer, first because of the white patches on its soft, fern-like leaves, then by its early white flowers and finally by its long, obvious, green fruits. It is also delightfully scented.

It grows better in the north but can easily be cultivated from seed in the south. Even where it occurs frequently and is naturalized, however, the plant is thought to have escaped originally from old monastery gardens.

According to Culpeper, Sweet Cicely 'groweth like Hemlock, but of a fresher green colour tasting as sweet as the aniseed'. He mentions too its 'pleasantness in salads' and the fact that 'it is so harmless you cannot use it amiss'. Culpeper recommended the roots 'to be boil'd like parsnip'.

Sweet Cicely

Sweet Cicely tea made from the crushed fruits is good for 'relieving wind in the belly' and 'for the tiresome cough' and is also a mild diuretic. The seeds can be chewed, like those of Angelica and Fennel.

The leaves, stems, flowers and fruits have synergic properties and act as a catalyst, sweetening acid fruits or rhubarb when cooked with them, which makes the plant a sugar-saver and good for slimmers.

It is also used in the preparation of different liqueurs, including Chartreuse.

SWEET FLAG   *Acorus calamus*

Country name: SWEET SEDGE.

Cardinal Wolsey brought Sweet Flags from the Norfolk Broads to strew the floors of his London house. The plant grows by water and is usually a relic of cultivated plants brought to Britain from south-east Asia during the sixteenth century.

Sweet Flag is also well-named Sweet Sedge, because its long, narrow leaves are easily mistaken for those of tall sedges as well as those of the Yellow Iris or Flag. The difference can be told directly you bruise a leaf and release the strong 'Citronella' smell. It can also be seen in the flowers which branch out of the leaves and look like green Arum spadices, or the spadix in the centre of Cuckoo-pint (Wild Arum).

Sweet Flag

A few leaves of this plant crushed and rubbed on hands, arms and bare legs, as well as on the forehead, work very well as fly-repellents. Medicinally, herbalists use distilled Sweet Flag in cases of chronic dyspepsia as a stimulant for anorexia-sufferers to promote their poor appetites. It is now also thought to have antibiotic properties.

But BEWARE, for this is not a plant to be tried out at home, apart from applying its juices externally to the skin as insect-repellents, because if used in incorrect dosage it can have toxic effects.

TANSY   *Tanacetum vulgare*

Country names: BATCHELOR'S-BUTTONS; BITTER or YELLOW BUTTONS; GINGER-PLANT; STINKING WILLIE.

Tansy, with its tall and handsome clusters of flat, yellow 'button' flowers and feathery leaves, has often cropped up in country-lore and has historical associations. It was one of the Pasque or Easter herbs from which 'Tansies' were prepared in memory of the bitter herbs taken by the Jews at the Passover, and the leaves of Tansy were also used as an ordinary spring tonic. William Coles suggested that it was eaten 'to counteract the saltiness of the large quantities of the fish consumed during Lent', but Victorian countrywomen made Tansy Puddings that sound mouth-watering from '$\frac{1}{4}$lb blanched almonds, pounded and placed in a stewpan to which a gill of the syrup of Roses, the inside of a fresh white bread roll, a pinch of grated Nutmeg, 3oz of butter, $\frac{1}{2}$-glass of Brandy, some slices of Lemon-peel' were all added to a tablespoonful of Tansy juice. The pudding was completed by pouring 'over the mixture 1$\frac{1}{2}$ pints of creamy milk with 8 eggs beaten into it and then bake carefully until set'.

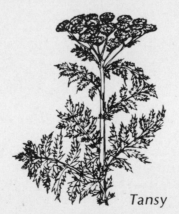

Tansy

Culpeper extols the medicinal virtues of Tansy: 'Now Dame Venus hath given women two herbs of one name, one to help conception, the other to maintain beauty; and what more can be expected of her. What now remains for you but to love your husband, and not to be wanting to your poor neighbours'!

Wild Tansy, he says, also 'Stayeth spilling or vomiting of the blood' and 'the same boiled with vinegar with honey and alum and gargled in the mouth, easeth the pains of the toothache, fasteneth loose teeth, helpeth the gums that are sore and settleth the palate of the mouth into its place ...'.

Modern herbalists still use this plant but they stress that it must only be taken in very small doses.

TAWNY GRISETTE   *Amanita fulva*

Although this graceful, thin and long-stemmed fungus is one of the *Amanita* genus which also contains the lethal Death-cap, the 'Destroying Angel' *and* the well-known scarlet-capped Fly Agaric, Tawny Grisettes, or *Amanitopsis* as they used to be called, are edible.

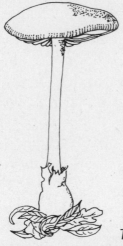

Tawny Grisette

They appear in early autumn and have tawny, rather tan-coloured shiny tops to their caps which are edged with neat striations. The gills are white and this fungus grows up out of a nearly-white, enclosing, soft-shelled, egg-like sheath called the *volva*. *It has no ring round the stem.*

Tawny Grisettes should be picked while young and eaten fresh and be cooked in butter, in a casserole, or be lightly fried.

THYME   *Thymus serpyllum*

The many varieties of Thyme used as a culinary herb are now known widely and grown in all kinds of containers by the gardenless, as well as adding to the fragrance of cottage and castle herb gardens.

Wild Thyme grows on remaining unploughed chalk hills and in other dry places in the countryside and was, of course, the ancestor of the many cultivated types.

When sheep roamed on the Downs and grazed all over their extensive pasture, the fragrance and flavour of the wild Thyme, Marjoram, Basil and other herbs were said to be responsible for the wonderful flavour of Southdown mutton and lamb. Then the constant grazing kept the turf very short and other plants that were pleasing to country-lovers, such as wild orchids, also had a chance to flourish.

Now that the sheep have been replaced by acres and acres of treeless, turfless fields, their scant top-soil fertilized with artificials and sown with crops of endless barley, it is difficult to find odd untouched corners where wild Thyme can be picked. If you get a few sprigs, make a cup of wild Thyme tea. It is refreshing and redolent of hot summer weather. Or dry it and keep it among your clothes in a drawer.

Thyme crushed and put into a 'little bag under the pillow encourages sleep for they who be short of it'. Indeed, Thyme tea is drunk at night 'to prevent the nightmare and other bad dreams'.

Thyme is still one of the plants put into herbal posies like that of the Queen on Maundy Thursday, or into Judges' nosegays. It has historical association as a fragrant antiseptic and was carried as such during epidemics of such diseases as the Plague, as well as to ward off 'all kindes of evil sperytes'.

Thyme is one of the herbs put into 'bouquets garnis' and its culinary properties are endless. Bees love the plant and their honey from it is said to be a good heart tonic.

TURNIP   *Brassica rapa*

Sometimes farmers leave corners of turnip fields unharvested, or odd Turnip plants become mere 'escapes' from cultivation and it is then worth gathering their tops for green vegetables, as well as their roots.

Turnips are recognizable by their bright green, slightly hairy and unbloomed leaves which spread out from the root without first ascending, as do some other farm-cultivated *Brassicae*, in a stout central stem. Turnip flowers are yellow and similar to those of Cabbages, but they open out *over* the unopened buds, which, if you look carefully at the flowers of other members of the Cabbage or Cruciferous family, you will find is unusual.

VIOLET   *Viola odorata*

Violet flowers, preserved or candied or crystallized, can be used as sweets, or to decorate cakes, puddings or even biscuits. The process of candying these and other innocuous coloured flowers like the cobalt stars of **Borage, Primroses, Cowslips,** petals of border **Carnations,** or wild **Angelica** is the same and to the undomesticated less difficult than it looks!

You will need a 'sweet' thermometer. Collect half a cupful of flowers and dry them. Melt 1lb (450g) castor sugar in a teacupful of water and bring it gently to the boil. While this is happening, warm up the sweet thermometer, put it into the sugary solution and watch for it to reach just over 240°F (115°C) and maintain this temperature for at least 60 seconds. Keep the thermometer in the pan and drop a few flowers – not more than 7 to 11 at a time – into the boiling syrup for one minute, watching the temperature which should stay at 240°. Lift the flowers out carefully with a sieve-spoon and put them down very lightly on to a sheet of foil. Replace them with others, till all are 'sugared'. Then lift the foil and place it in a pre-warmed oven *with no heat on,* and leave the flowers to dry.

If you are brave enough to turn the flowers delicately over from one side to the other once during their drying, it prevents distortion.

Uncandied, fresh and dry Violets have respiratory and

purgative medicinal properties and may also be given to children (see Primrose page 96) very sparingly in sandwiches, if they will not eat vegetables or salad plants.

Violet leaves have been credited by old country herb-women as being helpful for external breast cancers, if made into poultices.

WALNUT   *Juglans regia*

Walnuts are sought by rooks before they are ripe, plucked off and carried away from the trees to nearby fields and lawns to be skinned of the vitamin C-containing green 'husks', before being neatly opened and their kernels demolished.

If you can pick some Walnuts before the birds have them all, and there seems to be nothing to stop you shying sticks at the tree (if it is not on private ground) in July, according to the old English rhyme which says

A dog, a woman, a walnut-tree,
The more you beat 'em, the better they be!

the young walnuts can be pickled in their green outer covering, or Walnut Ketchup can be made. For this you will need some good jars. Cover the Walnuts, once you have almost filled a jar, with vinegar and tie a cover over it. Leave it for a year. After that, strain out the 'liquor', putting the Walnuts on one side, add 2 cloves of garlic, ½lb (225g) anchovies, 2pt (1·1l) red wine and 1oz (25g) mace, cloves, and ginger and plenty of pepper and then boil until the liquid is reduced to half the original quantity. Bottle it up the next day and use it as a fish-sauce. 'The longer it is kept, the better it is!' The Walnuts can be used as extra 'pickles'.

Fresh Walnuts need no description. Once you have peeled their outer 'husks' off, you will need no further telling that they provide a good dye or stain. Walnuts still in their shells can be buried, like Hazelnuts, in tight-lidded, sealed tins, for winter use.

WATERCRESS   *Nasturtium officinale*

Country names: TANG-TONGUES; KERSE; BILDERS.

It is a pity that Watercress-lovers have to be so careful

nowadays, for this old favourite and valuable free herb can no longer safely be picked wherever you find it, although there is still plenty about.

But BEWARE, for there are two sources of danger. *One,* Watercress may be infected with liver-fluke parasites which live as encysted larvae for part of their lives in some species of pond snails, and *two,* many of our best Watercress streams and dykes are now contaminated by infiltrating chemical toxic sprays which have seeped in from adjacent land, or, alternatively, by pollution from sewage.

Watercress

If you do find a clean site, Watercress makes a splendid salad plant and sandwich-filler. It is rich in vitamins A, C and E. It can be used in soups, too, but the high vitamin content will then be reduced by the cooking.

Watercress grown specially and commercially on clean, gravel-bottomed beds with a filtered stream running through them is safer than any except that picked from really remote, unpeopled and unfarmed places now.

WILD STRAWBERRIES   *Fragaria vesca*

There is a long history of gourmets transplanting wild Strawberry plants to their gardens or places where they could be manured and encouraged to grow bigger! This is not surprising for Wild Strawberries, although they often grow profusely, are very small and very low on the ground and therefore arduous to pick, even though their flavour is unique and delightful.

I think that they are worth the effort! They need no sugar and are perfect eaten with cream. It seems a sacrilege to cook them, but they make the most delicious jam that home-cooking can ever produce, although lemon-juice or a few green apples are needed to help the jam to set firmly.

It is no good counting on finding Wild Strawberries in the same place year after year. They flourish where the ground has been cleared and do well as long as they get plenty of light, but directly taller plants colonize the ground, the Strawberry plants have to struggle for light and their fruit yield becomes poor.

Herbalists say that a tea made from young Wild Strawberry leaves is every bit as good as China tea. The leaves should be picked in the spring and dried carefully and then stored in tightly-lidded jars until used. This tea has the reputation of being a good tonic for convalescents as well as being useful as an alkalizer for those whose systems are too acid.

WOOD SORREL   *Oxalis acetosella*

Country names: HALLELUJAH; EASTER-BELLS; EASTER SHAMROCK; CUCKOO'S-MEAT; CUCKOO'S-CLOVER; CUCKOO'S-BREAD AND CHEESE; BUTTER AND EGGS; GOD'S ALMIGHTY BREAD; RABBIT'S-MEAT; FOX'S-MEAT; SOUR; WHITSUN-FLOWER.

It is understandable that this acid-green, clover-leaf-shaped woodland plant should have earned such a wide variety of popular local names through the ages, for it is one of our prettiest woodland flowers. The list above is only a selection of the large total number known.

In early spring the folded leaves appear on bare woodland floors. They expand later but still stay lime-green and are foils for the pale pink, very delicately veined flowers which, in the south, are usually out to greet the popular cuckoo.

An odd leaf or two can be picked and eaten by country wanderers as thirst-quenchers and a few, if their taste has been approved, can be taken home to put in salads.

Wood Sorrel

Wood Sorrel has always been regarded as a magic three-leaved plant and has often been picked as a substitute for, or in mistake for, Shamrock, which is really one of the clovers.

YARROW   *Achillea millefolium*

Country names: MILFOIL; CAMMOCK; MELANCHOLY; STAUNCHWEED; NOSEBLEED; SNAKE'S-GRASS; TRAVELLER'S-EASE; WOUNDWORT.

Yarrow's virtues as a free, plentiful, ubiquitous weed lie chiefly in its medicinal properties, for which it has been known through the ages.

The feathery leaves can, though, be picked and chopped raw in salads to give them a weird, unusual tang and they can also be picked when young, and boiled until tender to make a green vegetable.

It is as a 'simple', or useful herb for home remedies, that the history of the uses of Yarrow could fill a book by themselves. Some of the plant's properties are highlighted in its country names. The ancients used it as a wound-healing, blood-staunching herb: Achilles was said to have saved the lives of thousands of his soldiers by using it on their wounds.

But 'Nosebleed', one of its Sussex names, may be partly derisory as although the plant was probably used to arrest nost-bleeding, leaves were also picked, rolled up dry and inserted into the nostrils of country maidens, to make them bleed! This was all part of the fun and games indulged in to prognosticate 'whether or not they could soon expect to meet their true loves'! It does, however, illustrate the dangers in trusting in popular plant names without knowing enough about them.

Infusions made from the whole Yarrow plant had the reputation of making first-rate foot-baths 'for those who have got weary on the way'. Ale made from Yarrow doubtless alleviated the weariness of travellers too.

It is strange that there seems to be no English country name for this herb's bitterness and for the fact that dairy cows which have grazed upon it to excess give milk that makes bitter butter.

Yarrow

But Yarrow tea, in all seriousness, is an excellent drink for anyone who wants to sweat out a cold, or who needs a digestive tonic. It is made in the usual way by washing about 1oz (25g) of the herb, putting it in a big jug and pouring 1pt (550ml) of boiling water over it and letting it stand for a few minutes before draining the liquid off to be drunk a wineglassful at a time, at intervals during the day.

An ointment made by boiling the whole, washed plant of Yarrow in lard until the fat turns green, when the herb should be sieved out and the green lard set aside in small pots to cool, is said to be a good hair-restorer. Culpeper says that 'it stayeth the shedding of the hair', and other old herbalists endorse this by adding that 'it cures ye baldnesse'.

THERAPEUTIC INDEX

INDEX